Applications of Advanced Nanomaterials in Water Treatment

Water is the most vital substance in every aspect of life, and its contamination because of the activity of mankind poses a big global challenge. Addressing this issue for drinking purposes and environmental protection, is the current big issue. Many research groups worldwide have been working on effective treatment technologies based on nanomaterials during the last two decades. Water and wastewater treatment by nanomaterial-based technologies has become an aid in finding possible solutions for contaminated wastewater released from various water sources. Nanoscale materials can be seen to take on unique and unpredictable properties that make them more robust, flexible, lighter, and faster, and the particular material used for the development of devices and systems.

Features:

- This book presents the use of different advanced nanomaterials to treat water and wastewater.
- The uses of carbon-based nanomaterials, metal-organic frameworks (MOFs), and biopolymer-supported nanomaterials for water treatment are explored, focusing on new-generation materials.
- Water purification methods such as the disinfection of water using green synthesized nanomaterials, adsorption through nanomaterial-based adsorbents, and nanofiltration techniques are discussed. We emphasize efficient water treatment methods and the use of new emerging nanomaterials in the toxicological study of nanomaterials.

Nanomaterials provide high-performance and cost-effective treatment for water treatment. Various nanomaterials and electrochemical methods are used to stop, remove, or neutralize water harmful organic and inorganic contaminants in water through adsorption, filtration, and disinfection.

Applications of Advanced Nanomaterials in Water Treatment

Edited by
Dinesh Kumar and Meena Nemiwal

CRC Press
Taylor & Francis Group
Boca Raton London New York

CRC Press is an imprint of the
Taylor & Francis Group, an **informa** business

First edition published 2023
by CRC Press
6000 Broken Sound Parkway NW, Suite 300, Boca Raton, FL 33487-2742

and by CRC Press
4 Park Square, Milton Park, Abingdon, Oxon, OX14 4RN

© 2023 selection and editorial matter, Dinesh Kumar and Meena Nemiwal; individual chapters, the contributors

CRC Press is an imprint of Taylor & Francis Group, LLC

Reasonable efforts have been made to publish reliable data and information, but the author and publisher cannot assume responsibility for the validity of all materials or the consequences of their use. The authors and publishers have attempted to trace the copyright holders of all material reproduced in this publication and apologize to copyright holders if permission to publish in this form has not been obtained. If any copyright material has not been acknowledged, please write and let us know so we may rectify in any future reprint.

Except as permitted under U.S. Copyright Law, no part of this book may be reprinted, reproduced, transmitted, or utilized in any form by any electronic, mechanical, or other means, now known or hereafter invented, including photocopying, microfilming, and recording, or in any information storage or retrieval system, without written permission from the publishers.

For permission to photocopy or use material electronically from this work, access www.copyright.com or contact the Copyright Clearance Center, Inc. (CCC), 222 Rosewood Drive, Danvers, MA 01923, 978-750-8400. For works that are not available on CCC please contact mpkbookspermissions@tandf.co.uk

Trademark notice: Product or corporate names may be trademarks or registered trademarks and are used only for identification and explanation without intent to infringe.

Library of Congress Cataloging-in-Publication Data
Names: Kumar, Dinesh (Chemist), editor. | Nemiwal, Meena, editor.
Title: Applications of advanced nanomaterials in water treatment /
edited by Dinesh Kumar and Meena Nemiwal.
Description: First edition. | Boca Raton : CRC Press, [2023] |
Includes bibliographical references and index.
Identifiers: LCCN 2022025607 (print) | LCCN 2022025608 (ebook) |
ISBN 9781032181165 (hb) | ISBN 9781032181172 (pb) | ISBN 9781003252931 (eb)
Subjects: LCSH: Water–Purification–Materials. | Nanostructured materials.
Classification: LCC TD477 .A67 2023 (print) |
LCC TD477 (ebook) | DDC 628.1/62–dc23/eng/20220819
LC record available at https://lccn.loc.gov/2022025607
LC ebook record available at https://lccn.loc.gov/2022025608

ISBN: 978-1-032-18116-5 (hbk)
ISBN: 978-1-032-18117-2 (pbk)
ISBN: 978-1-003-25293-1 (ebk)

DOI: 10.1201/9781003252931

Typeset in Times New Roman
by Newgen Publishing UK

Contents

Preface vii
About the Editors ix
Contributors xi
List of Abbreviations xiii

Chapter 1 Introduction: Nanomaterials and Their Role in
Water Treatment 1

Swati Joshi and Meena Nemiwal

Chapter 2 Advanced Carbon-Based Nanomaterials for Treatment
of Water 25

*Sapna Nehra, Sayed Zenab Hasan, Rekha Sharma,
and Dinesh Kumar*

Chapter 3 Metal-Organic Frameworks (MOFs) for Water Treatment 45

Manjula Nair and Kajal Panchal

Chapter 4 Biopolymers Supported Nanomaterials for Water Treatment 63

*Pragati Chauhan, Sapna Nehra, Rekha Sharma,
and Dinesh Kumar*

Chapter 5 Disinfection of Water by Eco-friendly Nanomaterials 81

Poonam Ojha and Meena Nemiwal

Chapter 6 Advance Nanomaterials-Based Adsorbents for Removal
and Recovery of Metals from Water 97

W. M. Dimuthu Nilmini Wijeyaratne

Chapter 7 Photocatalytic Degradation of Organic Pollutants by
Using Efficient Nanomaterials 125

*Vijayalakshmi Gosu, Gayatri Rajpur, Uttam Singh,
Meena Nemiwal, S. Suresh, and Verraboina Subbaramaiah*

Chapter 8 Recent Development in Nanofiltration Membrane for
Water Purification 147

*Sayed Zenab Hasan, Sapna Nehra, Rekha Sharma, and
Dinesh Kumar*

Chapter 9 Nanomaterials for Electrochemical Treatment of
Pollutants in Water 165

Chetan Kumar and Ritu Painuli

Chapter 10 Nanotoxicology and Challenges of Using Nanomaterials
for Water Treatment 185

Nitin Kumar Sharma and Jyotsna Vishwakarma

Index 199

Preface

> "We forget that Water cycle and Life Cycle are one"
> —Jacques Yves Cousteau

Water is vital for every aspect of life, and its contamination by pollutants poses a significant global challenge. Its treatment for drinking purposes and environmental protection is the current big issue. In the last two decades, many research groups worldwide have been working on effective treatment technologies based on nanomaterials. Water and wastewater treatment by nanomaterial-based technologies has become an aid in finding possible solutions for contaminated wastewater released from various water sources. Nanoscale materials are seen to take on unique and hitherto unexpected properties that make them more robust, flexible, lighter, and faster, and the ideal material to be used for the formation of devices and systems. Nanotechnology produces high-performance and cost-effective equipment for water treatment, and nanomaterials also provide on-spot and continuous monitoring of water quality. Various nanomaterials and electrochemical methods are used to relent, remove, or neutralize water harmful organic and inorganic contaminants of water that are dangerous to the productivity and integrity of ecosystem/human health.

This book presents the use of different nanomaterials for the treatment of water and wastewater. This book will help fill a gap in understanding the nanotechnology and electrochemical methods used for water treatment. Chapter 1 presents a brief introduction to advanced nanomaterials and their role in water treatment. Chapter 2 deals with the use of carbon-based nanomaterials for water treatment with emphasis on new-generation materials. Chapter 3 describes the different types of metal-organic frameworks (MOFs)-based nanomaterials used to treat wastewater. Chapter 4 explores the advancement in biopolymer-supported nanomaterials for water purification. Chapter 5 gives the details of disinfection methods using green synthesized nanomaterials to remove pathogenic micro-organisms to purify water. Chapter 6 discusses the treatment of metals in water by employing nanomaterial-based adsorbents. Chapter 7 focuses on the photocatalytic treatment of water by exploiting efficient nanomaterials. Chapter 8 deals with the purification of water by using the nanofiltration technique. Chapter 9 reveals the role of nanomaterials for water treatment by electrochemical methods. Chapter 10 is about the toxicology of nanomaterials and the challenges in using them to treat water.

About the Editors

Dinesh Kumar is currently working as Professor in the School of Chemical Sciences at the Central University of Gujarat, Gandhinagar, India. Prof. Kumar obtained his master's and Ph.D. degrees in Chemistry from the Department of Chemistry, University of Rajasthan, Jaipur. Prof. Kumar has received many national and international awards and fellowships. His research interest focuses on the development of capped MNPs, core-shell NPs, and biopolymer incorporated metal oxide-based nanoadsorbents and nanosensors for the removal and the sensing of health hazardous inorganic toxicants like fluoride and heavy metal ions from aqueous media. Prof. Kumar developed hybrid nanomaterials from different biopolymers like pectin, chitin, cellulose, and chitosan for water purification. His research interests also focus on the synthesis of supramolecular metal complexes, metal chelates, and their biological effectiveness. He has authored and co-authored over 122 publications in journals of international repute, three books, over eight dozen book chapters, and 80 presentations/talks at national/international conferences.

Meena Nemiwal is presently Assistant Professor in the Department of Chemistry, Malaviya National Institute of Technology, Jaipur, Rajasthan, India. She obtained her master's and Ph.D. degrees from the Department of Chemistry, University of Rajasthan, Jaipur in 2002 and 2007 respectively. She qualified as NET-JRF(CSIR) in 2005 and GATE-2007. She has been in teaching for ten years. She wrote her Ph.D. thesis, entitled "Electrochemical Studies of Metal Ligand Complexes at Dropping Mercury Electrode in Aqueous and Aqueous Non- Aqueous Media". Her research work was on the study of toxic heavy metals such as lead and cadmium and the stability of their complexes with different ligands. Her research interest focuses on the synthesis of functionalized nanoparticles, nanostructured materials, and their application for electrode modification in the electrochemical sensing of metal ions in water. Dr. Meena has authored several research articles/reviews/book chapters in reputed publications. She has attended and presented papers in many National and International conferences. She is a lifetime member of the academic societies: The Indian Science Congress and The Indian Desalination Association. She has delivered many talks and has organised several faculty development programmes, National and International conferences.

Contributors

Pragati Chauhan
Department of Chemistry, Banasthali Vidyapith, Rajasthan, India

Vijayalakshmi Gosu
Department of Chemical Engineering, Malaviya National Institute of Technology Jaipur, Jaipur, India

Sayed Zenab Hasan
Department of Chemistry, Dr. K. N. Modi University, Newai, Rajasthan, India

Swati Joshi
Department of Chemistry, Swami Keshvanand Institute of Technology, Management and Gramothan, Jaipur, India

Chetan Kumar
Natural Product Chemistry Division, CSIR–Indian Institute of Integrative Medicine, Canal Road, Jammu, India

Dinesh Kumar
School of Chemical Sciences, Central University of Gujarat, Gandhinagar, Gujarat, India

Manjula Nair
Heriot Watt University, Academic City, Dubai

Sapna Nehra
Department of Chemistry, Nirwan University, Jaipur, Rajasthan, India

Meena Nemiwal
Department of Chemistry, Malaviya National Institute of Technology, Jaipur, Rajasthan, India

Poonam Ojha
Department of Chemistry, Swami Keshvanand Institute of Technology, Management and Gramothan, Jaipur, India

Ritu Painuli
Department of Chemistry, Banasthali Vidyapith, Rajasthan, India

Kajal Panchal
School of Chemical Sciences, Central University of Gujarat, Gandhinagar, Gujarat, India

Gayatri Rajpur
Department of Chemical Engineering, Malaviya National Institute of Technology Jaipur, Jaipur, India

Nitin Kumar Sharma
Department of Chemical Engineering, Indian Institute of Technology, Kanpur, India and Shri Maneklal M. Patel Institute of Sciences and Research, Kadi Sarva Vishwavidyalaya, Gandhinagar, Gujarat, India

Rekha Sharma
Department of Chemistry, Banasthali Vidyapith, Rajasthan, India

Uttam Singh
Department of Chemical Engineering,
Malaviya National Institute of
Technology Jaipur, Jaipur, India

Verraboina Subbaramaiah
Department of Chemical Engineering,
Malaviya National Institute of
Technology Jaipur, Jaipur, India

S. Suresh
Department of Chemical Engineering,
Maulana Azad National Institute of
Technology, Bhopal, India

Jyotsna Vishwakarma
K.B. Pharmacy Institute of Education
and Research, Kadi Sarva
Vishwavidyalaya, Gandhinagar,
Gujarat, India

W. M. Dimuthu Nilmini Wijeyaratne
Department of Zoology and
Environmental Management,
Faculty of Science, University of
Kelaniya, Dalugama, Kelaniya,
Sri Lanka

Abbreviations

$[C_4Bth][PF_6]$	Benzothiazole Ionic Liquid (*N*-butyl benzothiazole hexafluoroborate)
0D	Zero Dimensional
1D	One Dimensional
2D	Two Dimensional
3D	Three Dimensional
8YSZ	Yttria–Stabilized ZrO_2 (8 mol%)
AgNPs	Silver Nanoparticles
AM and BB	Amaranth and Brilliant Blue
ANT	Anthracene
AOP	Advanced Oxidation Processes
AR	Aerogels
AuNPs	Gold Nanoparticles
BZT	Benzotriazole
CCD	Central composite design
CNCs	Cellulose Nanocrystals
CNFs	Cellulose Nanofibrils
CNTs	Carbon Nanotubes
COD	Chemical Oxygen Demand
COFs	Covalent Organic Frameworks
CoPc	Cobalt Phthalocyanine
CS	Chitosan
CUR	Carbon Usage Rate
CVD	Vapor Deposition Method
CWFs	Ceramic Water Filters
EBCT	Empty Bed Contact Time
ED	Ethylenediamine
ED–g–MWCNT	Ethylene Diamine Joined Multiwalled Carbon Nanotubes
EPA	Environmental Protection Agency
EY	Eosin yellow
FESEM	Field Emission Scanning Electron Microscopy
FRR	Flux Recovery Rate
f-SWCNTs	Functionalized Single-Walled Carbon Nanotubes
f-SWCNTs	Functionalized Single-Walled Carbon Nanotubes
FTIR	Fourier Transform Infrared Spectroscopy
GG	Guar-Gum
GO	Graphite oxide
IL-MWCNTs	Liquid-multi walled carbon nanotubes
IP	Interfacial Polymerization
MB	Methylene blue

MF	Microfiltration
MG	Malachite Green
MMMs	Mixed Matrix Multimembranes
MOF	Metal–Organic Framework
MONPs	Metal Oxide Nanoparticles
MPs	Microplastics
MTV-MOF	Multivariate Amino Acid-Based Metal–Organic Framework
MXene@CA	Crosslinked–Ti_3C_2Tx(MXene)/Cellulose acetate
NF	Nanofiltration
NOM	Natural organic matter
NPs	Nanoparticles
O–POPs	O–hydroxyazo porous organic polymers
PA	Polyamide
PAH	Poly Allylamine Hydrochloride
PDADMAC	Poly–Dialllyl–Dimethyl–Ammonium–Chloride
PDDA	Polydiallyldimethylammonium chloride
PE	Polyethylene
PEM	Polyelectrolyte Multilayer
PES	Polyethersulfone
PET	Polyethylene terephthalate
PhACs	Pharmaceutically Active Compounds
PIP	Piperazine
PMS	Peroxymonosulfate
POSS	Polyhedral Oligosilsesquioxane
PSf	Polysulfone
PSO	Pseudo-Second Order
PSS	Polysodium–4–styrenesulfonate
PVP	poly(vinylpyrrolidone)
QCBNPs	Quaternized Carbon–Based Nanoparticles
Rh6B	Rhodamine Blue
RO	Reverse Osmosis
ROS	Reactive Oxygen Species
SEDFRT	Solution–Electro–Diffusion–Film Integrated with Reactive Transport
SEM	Scanning Electron Microscopy
SWCNT-BP	Single-Walled Carbon Nanotube Buckypaper
TC	Tetracycline
TEM	Transmission Election Microscopy
TFC	Thin Film Composite
TFN	Thin Film Nanocomposite
TGA	Thermogravimetric Analysis
TiO_2 NPs	Titanium dioxide Nanoparticles
TOC	Total Organic Carbon

List of Abbreviations

TSS	Total Suspended Solids
UF	Ultrafiltration
UNICEF	United Nations International Children's Emergency Fund
VSM	Vibrating Sample Magnetometer
WHO	World Health Organization
XPS	X–Rays Photoelectron Spectroscopy
XRD	X–Rays Diffraction
ZnO NPs	Zinc Oxide Nanoparticles
ZVI	Zero-valent Iron NPs
MFC	Magnetic Framework Composites
MDC	MOF-Derived Carbon
EOC	Emerging Organic Contaminants

1 Introduction
Nanomaterials and Their Role in Water Treatment

Swati Joshi[1] and Meena Nemiwal[2*]
[1]Department of Chemistry, Swami Keshvanand Institute of Technology, Management and Gramothan, Jaipur, India 302017
[2]Department of Chemistry, Malaviya National Institute of Technology, Jaipur, India 302017
*Corresponding author: Email: meena.chy@mnit.ac.in

CONTENTS

1.1 Introduction 2
1.2 Classification of Nanomaterials 3
    1.2.1 Classification of Nanomaterials Based on Particle Dimensions 4
    1.2.2 Classification of Nanomaterials Based on Pore Dimensions 4
    1.2.3 Classification of Nanomaterials Based on Construction 5
1.3 Nanomaterials for Water Remediation 5
    1.3.1 Zero-Valent Metal NPs 6
        1.3.1.1 Silver Nanoparticles 6
        1.3.1.2 Iron Nanoparticles 7
    1.3.2 Metal Oxide Nanoparticles 8
        1.3.2.1 TiO_2 Nanoparticles 8
        1.3.2.2 ZnO Nanoparticles 9
        1.3.2.3 Iron Oxide Nanoparticles 9
    1.3.3 Carbon-Based Nanomaterials 10
    1.3.4 Nanocomposites 11
1.4 Water Treatment Methodologies 12
    1.4.1 Nanophotocatalysts 12
    1.4.2 Nanomembranes 14
    1.4.3 Nanosorbents 15
1.5 Conclusion 16
References 17

1.1 INTRODUCTION

Nanomaterials are usually defined as materials in which at least one dimension of the structural components is sized between 1 and 100 nm [1]. The overall bulk composition and the actual component nanomaterial show similarity, but might differ in terms of their physicochemical properties. The mechanical, electrical, optical, and magnetic properties of nanomaterials differ from conventional materials because of nano-size and large surface area. Nano-sized materials sometimes show drastic differences in properties compared to macro-sized particles [2]. For example, macro-sized Cu is opaque while its nano-size is transparent, macro-sized Pt is inert whereas its nano-size shows catalytic properties, larger-sized Al is stable whereas its nano product is combustible, and gold is solid at room temperature (RT), but its nano-size becomes liquid at RT. There has been noticeable research and development in nanomaterials because of their peculiar properties and applicability in diverse fields such as medicine, biology, engineering, catalysis, sensing, and the like [3–7].

The production of nanomaterials requires one of two alternative approaches, this could be the bottom-up or the top-down approach [8]. In the bottom-up method, preparation is carried out by an atom on atom or molecule on molecule method. In contrast, smaller materials are framed from larger structures in the top-down method.

Nanomaterials can be realized as intermediary structures between single atoms and bulk materials. The major difference between bulk materials and nanomaterials arises in of the larger surface areas and quantum effects [9]. These properties alter or modify some of the properties of nanomaterials such as electrical, mechanical, chemical reactivity, and strength to a more significant extent [10]. As the particle size is reduced, more atoms come to the surface. All the catalytical activities are performed through surface reactions, so the nanoparticles (NPs) show a high rate of chemical reactivity compared to their respective bulk compounds because the nano-sized materials can potentially have a million times higher surface areas. Nanomaterials derived from new emerging materials such as metal-organic frameworks (MOFs) [11–14] and covalent organic frameworks (COFs) [15] have been exploited in a wide range of applications such as water purification, fuel cells, drug delivery, sensing, catalysis, and so forth.

Water is frequently contaminated by various factors such as industrial wastewater discharges, the discharge of untreated sewage water into water bodies, the discharge of wastes from agricultural sites, and the like [16]. Water remediation is a broader term comprising cleaning, sanitization, and restoring the damaged materials to make water suitable for health and palatable for drinking. Many technologies have been used so far for the decontamination of water, including chemical, physical, biological, and mechanical procedures. Using nanomaterials for water remediation has now drawn the attention of many researchers as these particles show many characteristic peculiarities such as nano-size, which increase the surface area of the particles and make them highly reactive and showing catalytic properties, moving tendencies in solutions,

Introduction: Nanomaterials and Their Role in Water Treatment

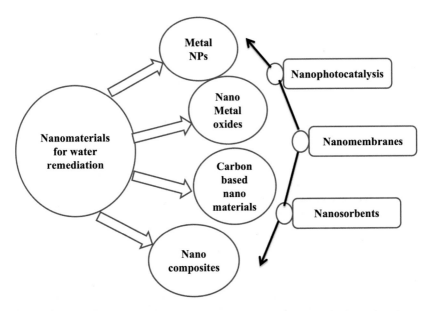

FIGURE 1.1 Schematic presentation of different nanomaterials and their applications for water remediation.

porosity, hydrophobicity, and hydrophilicity, strong mechanical properties, and so forth. [17]. Nanomaterials can serve the purpose of removing organic and inorganic pollutants, insecticides, pesticides, toxic chemicals, heavy metals, and pathogenic and non-pathogenic micro-organisms. Nowadays, with the development of research and advances in this field, nanomaterials are used in photocatalysts, membranes, sorbents, polymeric materials, and so forth, for the decontamination of polluted water (Figure 1.1). Introducing nanomaterials with classification of their roles in water treatment as photocatalysts, nanomembranes, and nanosorbents is discussed below.

1.2 CLASSIFICATION OF NANOMATERIALS

The properties of nanomaterials differ from their bulky counterpart to a large extent because of their high surface-area-to-volume ratio and quantum effects. Quantum effects are being exhibited by nanoscale materials. As these particles have number of electrons N which experience discrete energy levels because of quantum size. At nanoscale, quantum effects dominate the optical and electrical properties of the system. Nanoparticles have very small sizes but are larger than that of the quantum dots (QDs) sizes in the range of 8 - 100 nm. Thus, NPs shows properties between those of bulk and atoms or molecules.

According to the European Commission, nanomaterials represent a class of materials including elements, compounds, and composites in which at least one dimension is present in the range 1 - 100 nm. The following section of the chapter includes the classification of nanomaterials based on dimensions, pore dimensions, and construction.

1.2.1 Classification of Nanomaterials Based on Particle Dimensions

As far as dimensions are concerned, nanomaterials can be categorized into zero-dimensional (0D), one-dimensional (1D), two-dimensional (2D), and three-dimensional (3D) structures.

In 0D nanomaterials, none of the dimensions is larger than 100 nm, as in NPs and quantum dots. Quantum dots are nano-sized particles with semiconductor properties and generally with core-shell structure. Unique optical properties are exhibited by these quantum dots due to which light of specific wavelength is emitted upon energy application

Nanoparticles are abundantly found in nature as the products of photochemical reactions and released by plants and algae and in vehicle exhausts. Nanoparticles have been involved in cosmetics, paints, and targeted medical drugs delivery. Nano-sized semiconductor quantum dots can absorb or emit wavelengths in the visible region of light and have shown significant applications in biology and related studies.

Materials in which one of the dimensions is beyond the nanoscale are called 1D nanomaterials, for example, nanotubes, rods, and wires [18]. Carbon nanotubes are tubes of rolled graphene sheets [19]. They are quite strong mechanically, and their strength is comparable to that of a diamond. Similarly, oxide-based nanotubes such as TiO_2 have been used in catalysis, especially in photocatalysis and nanowires constructed from silicon, gallium nitride, and indium phosphides in data storage.

Plate-like nanomaterials or 2D nanomaterials, for example, graphene, nanofilms, coatings, and layers, have two dimensions out of the nanoscale. Graphene is carbonaceous with a polygonal lattice and has found applications in energy storage such as batteries, lightweight solar cells, and the like. [20].

3D nanomaterials [21] are not limited to nanoscale in any dimensions, and include bulk powders, nanowire bundles, nanoflowers, nanopillars, nanocoins, fullerenes, and the like. C_{60} fullerene molecules are spherical with 1 nm diameters, (maybe resembling a football) and are used for lubrication and electronic circuits.

1.2.2 Classification of Nanomaterials Based on Pore Dimensions

A popular classification method for nanoporous materials is based on the diameter size of their pores. Most of the important properties that play an important role in applications based on adsorption and diffusion characteristics are dependent on the pore diameter. The properties of the materials alter significantly during interactions with other molecules in the range of 1–100 nm.

It is important to note that the pore diameter determines the size of molecules that could penetrate or diffuse inside. The relative diameters of the guest molecule and

Introduction: Nanomaterials and Their Role in Water Treatment

pore size indicates interaction and diffusion properties. Based on pore dimension, IUPAC has classified nanoporous materials into three main groups:

Microporous materials (d < 2 nm) are nanomaterials with very narrow pores. Due to narrow pore sizes, only gases and small linear molecules can enter. These materials have found applications in the purification of gases and as membrane filters, for example, Na-Y nanomaterials and zeolites [22].

Mesoporous materials (2 < d < 50 nm) have comparatively larger molecules. The aromatics and some bulky monomers of the polymeric system can enter due to sufficiently large pore diameters [23]. These materials can be used for the adsorption of some liquids or vapors, for example, carbon mesoporous materials.

Macroporous systems (d > 50 nm) can host larger molecules such as macromolecules, polymers, small biopolymers, and the like [24]. These systems are used as catalytic systems and as sensing units because of the easy diffusion of certain chemicals in their pore systems, for example, carbon tubes, gels, and glasses.

1.2.3 Classification of Nanomaterials Based on Construction

Nanomaterials can be categorized into carbon-based composites, dendrimers, and metal-based nanomaterials.

Carbon-based nanomaterials have been found to have applications in various sectors in tubes, sheets, spheres, and NPs, for example, graphene, carbon nanotubes, and so forth.

Composites are nanomaterials combined with other nanomaterials or any bulk materials that might be organic or inorganic.

Dendrimers are multivalent, branched nano-sized polymeric materials formed via the self-assembly method. Their surfaces have many chain-ends. Also, 3D dendrimers have inner cavities in which other materials may reside or chemically react. Therefore, dendrimers can be used for water cleaning through metal ion trapping from the water that passes through them.

Metal-based nanomaterials include nano zero-valent metals and metallic oxides like Ag, Zn, Fe, TiO_2, FeO, ZnO, quantum dots, and so forth.

Various criteria for the classification of nanomaterials can be summarized in Figure 1.2.

1.3 NANOMATERIALS FOR WATER REMEDIATION

Nanomaterials have application in information technology, everyday life, health and hygiene, environmental protection, procuring transportation, energy sector, catalysis, and so forth. The peculiar and special properties of nanomaterials can be used to treat drinking water and wastewater. A study by Boschi-Pinto and co-workers [25] revealed that about 187 million child deaths are observed to occur simply due to unsafe drinking water every year. Ceramic water filters (CWFs) have become useful for treating water from disease-causing microbes or pathogens.

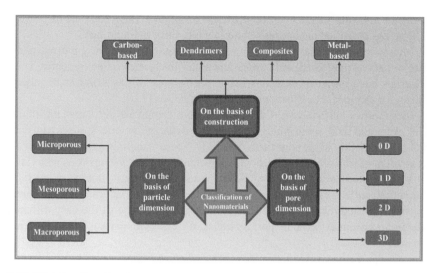

FIGURE 1.2 Classification of nanomaterials.

CWFs can also be combined with silver doped nanomaterials, making them suitable for removing pathogens. Carbon nanotubes, silver nanoparticles (AgNPs), fullerenes, and zeolites impregnated on mesoporous ceramic materials behave as sorbents and are used to remove water hardness. These nanomaterials have active properties against microbes, are chemically inert, and do not produce any harmful by-products after the treatment reaction. The catalytic action of metal oxide NPs with pollutants in which oxidation occurs degrades the pollutants. Water coming from agricultural and industrial sites contains many organic pollutants, and doped NPs are used for their removal.

1.3.1 Zero-Valent Metal NPs

This includes zero-valent Ag, Fe, and ZnNPs, that are the most extensively used materials in the modern era for water decontamination.

1.3.1.1 Silver Nanoparticles

AgNPs work effectively against many microbes present in drinking water. There are two explanations for the killing action of AgNPs on microbes. One of the explanations suggests that AgNPs adhere to the cell walls of microbes, causing severe damage to the cell walls and making them more porous [26]. A second explanation is that AgNPs generate free radicals in contact with microbes. These radicals destroy the cell membrane and thus kill the cells of micro-organisms [27]. AgNPs also attack S and P, combined with cell DNA. Cell enzymatic action is also deactivated by the –SH group of enzymes with Ag^+ ions released by the dissolution of AgNPs [28].

Introduction: Nanomaterials and Their Role in Water Treatment

A membrane is constructed by mixing AgNPs on ceramic filters with clay and sawdust for the decontamination of drinking water, especially removing *E. coli* easily [29]. Colloidal AgNPs can also be mixed with ceramic filters via the chemical reduction method. AgNPs were synthesized and impregnated into polyethersulfone (PES) membranes. These membranes were remarkably active against bacteria present in water [30]. There is also a drawback associated with using AgNPs like these because they accumulate in the water, and this causes a slow degradation of its properties.

1.3.1.2 Iron Nanoparticles

Zero-valent iron NPs (ZVI) also demand attention, showing remarkable water remediation properties. Their tiny size combined with a larger surface area makes the adsorption tendency of ZVI higher than other compounds. It also indicates reduction ability (with a reduction potential of $-0.440\,E^0/V$) and causes oxidation reactions in water. Further, its production and use costs are quite low due to its availability. FeNPs have been successfully examined for water pollution treatment due to their comparatively low cost, high adsorption tendency, precipitation, and oxidation ability. In aqueous media (with dissolved O_2), H_2O or H^+ might oxidize Fe(0), and as the result of a chemical reaction, Fe(II) and H_2 are formed, and these behave as reducing agents against water contaminants.

$$Fe^0 + 2H_2O \longrightarrow Fe^{2+} + H_2 + 2OH^-$$

$$Fe^0 + 2H^+ \longrightarrow Fe^{2+} + H_2$$

In slight alkaline media, a redox reaction occurs between Fe(II) and contaminants to form Fe(III), which is ultimately converted to a powerful flocculant, $Fe(OH)_3$, which is highly effective in the degradation of microbes [31]. Other than this, nZVI is also a powerful agent against various types of organic compounds in the presence of dissolved oxygen. An electron transfer reaction occurs to produce hydrogen peroxide upon the reaction of nZVI with D.O.

$$Fe^0 + O_2 + 2H^+ \longrightarrow Fe^{2+} + H_2O_2$$

$$Fe^{2+} + H_2O_2 \longrightarrow Fe^{3+} + HO^\cdot + OH^-$$

Recombination of Fe^{2+} with H_2O_2 (Fenton reaction) produces hydroxyl free radicals, a power oxidizing agent against many organic pollutants in the water. Owing to adsorption, reduction, and oxidation (in the presence of DO), nZVI has exhibited its action against a wide range of contaminants such as halogenated organic compounds, dyes, heavy metals, and many more [32].

However, nZVI particles are used as disinfectants in the water. There are many drawbacks such as aggregation, oxidation, and waste removal after the disinfection reaction. To get rid of these hurdles, the activity of nZVI particles can be modified by doping with any metal, emulsification, surface painting, and conjugation with support [33].

Despite the studies of FeNPs as potential agents for water remediation, ZnNPs have also attracted considerable attention. As far as standard reduction potential values are concerned, Zn shows a higher reducing ability than iron. The higher the reducing ability, the higher the rate of pollution degradation. Under suitable conditions, nZVZ can effectively degrade carbon tetrachloride in comparison to nano zero-valent iron particles. A comparative study was done to review the effects of various nano zero-valent metals on octachlorodibenzo-p-dioxin (OCDD). The results revealed that ZnNPs were significantly more active in OCDD dechlorination in comparison with zero-valent Al, Ni, and FeNPs [34]. The actions of ZnNPs are confined to degrading halogen-based organic pollutants only. Researchers are also in the process of checking the reactivity of nZVZ against other pollutants.

1.3.2 Metal Oxide Nanoparticles

Other than zero-valent metal cations, several metallic oxide NPs also have relevance in water remediation.

1.3.2.1 TiO_2 Nanoparticles

Out of many metallic oxides, researchers focused on TiO_2 due to its high-quality photocatalytic activity, lower cost, and stability in the chemical, physical, biological, thermal, and photochemical environment. Recently, it has been found that in the presence of light, metal oxides behave as catalysts and participate in the oxidative degradation of reducing substances in water, leading to the formation of intermediate compounds having low molecular masses and finally conversion into CO_2, H_2O and anions- nitrates, phosphates, and chlorides. The studies of Fujishima and co-workers on TiO_2 semiconductor electrodes in the electrochemical degradation of water in the presence of light led to the attention of various researchers on photocatalytic degradation [35].

In TiO_2 particles, UV radiation is required to induce charge separation corresponding to a high bandgap (3.2 eV). As a result of UV exposure, TiO_2 produces reactive oxygen species (ROS), which readily act upon contaminants and degrade them. TiO_2 NPs generate hydroxyl radicals under ultra-violet radiations, attacking cells of various microbes present in water and damaging them. TiO_2 NPs can be used to destroy any type of contaminants, such as compounds containing chlorine, dyes, pesticides, heavy metals, and so forth. [36,37]. Other than materials, the photocatalytic action of TiO_2 is very effective against a variety of microbes, including Gram positive and gram negative bacteria and viruses, algae, fungi, and protozoa [38].

However, because of the large bandgap, it needs UV rays for excitation. In the presence of visible radiation, the activities of TiO_2 are still unmarked. To increase the photocatalytic activity of TiO_2 under UV radiation and to enable TiO_2 to adsorb in the visible region, metal, and non-metal doping has been performed, which enhances the photocatalytic action of TiO_2 in the visible region against bacteria and viruses [39,40]. In addition, a UV radiation source is also required for the photocatalytic activity of TiO_2 NPs. Besides its reactivity, its preparation is also a tedious process, as is the very complicated recovery of TiO_2 NPs from the treated water. To solve this problem, TiO_2 particles have been impregnated with membranes like polyvinylidene fluoride and polymethylmethacrylate to facilitate its recovery [41].

1.3.2.2 ZnO Nanoparticles

Another metal oxide that is also extensively studied is ZnO NPs. The properties of ZnO NPs are very much similar to that of TiO_2, such as a wide-bandgap in the near UV spectral region, photochemical stability, and oxidizing tendency [42]. ZnO NPs are also considered to be eco-friendly due to their compatibility with various species. For these reasons, they are widely used to treat water. As with TiO_2 NPs, ZnO NPs also show absorption in the UV region due to high bandgap energies around 3.2 eV.

With the action of ZnO NPs, photo-corrosion may occur, which clogs the photocatalytic activity of these particles due to the recombination of photo-generated charges. Various anionic, cationic, rare earth, and co-dopants can be doped with ZnO NPs to increase the efficiency of ZnO NPs. Combining ZnO NPs with several semiconductors such as CdO, CeO_2, graphene oxide, and the like, has also been demonstrated to exhibit high-grade photodegradation [43].

1.3.2.3 Iron Oxide Nanoparticles

Because of the availability and ease of handling, iron oxide NPs can also be used for a wide range of applications, including catalysis and water treatment, especially for removing heavy metals present in water as contaminants [44]. Various forms of iron oxide nanoparticles are used as nanoadsorbents or nanosorbents. Nonmagnetic hematite (α-Fe_2O_3), magnetic magnetite (Fe_3O_4), and magnetic maghemite (γ-Fe_2O_4) are some common examples of such categories. As the nanosorbent particles are of nanoscale size, their separation and recovery from treated water is quite difficult. By applying an external magnetic field, magnetic nanosorbents like magnetite and maghemite can be recovered easily, and are used for removing heavy metals from water. Various types of ligands like EDTA, L-glutathione (GSH), meso-2,3-dimercaptosuccinic acid (DMSA) [45], and the copolymers of acrylic acid and crotonic acid [46] can be incorporated with iron NPs to improve their adsorption tendency and to discard the other interfering metal ions. Polymer molecules bind the metal ions present in water and behave as carriers. Nano hematites (α-Fe_2O_3) are found to show activity in sensing, catalyzing, and have reported environmental applications and for the removal of heavy metal ions from water. The 3D flower-like α-Fe_2O_3 provides a large surface area for the contaminants to react with. Apart from the larger surface area, the availability of multiple spaces and pores also made these 3D nanopetal subunits suitable for water treatment and they were reported to be the most suitable materials for the removal of As(V) and Cr(VI) [47]. Similarly,

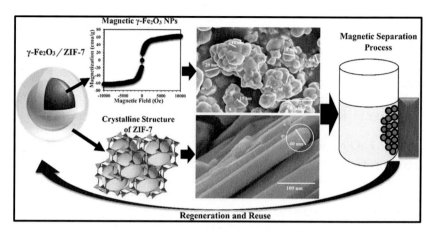

FIGURE 1.3 Composite of γ-Fe$_2$O$_3$/ZIF-7 for sorption of hydrocarbon pollutants in oily water ([48]. License: https://creativecommons.org/licenses/by/4.0/).

γ-Fe$_2$O$_3$/ZIF-7 structures were prepared with an enhanced surface area with no change in the crystal structure of γ-Fe$_2$O$_3$ NPs. These synthesized NPs were highly hydrophobic and exhibited good sorption properties for hydrocarbon pollutants in oily water (Figure 1.3) [48].

1.3.3 Carbon-Based Nanomaterials

Carbon nanomaterials hold unique structural peculiarities, electronic properties, and sorption capabilities. Their involvement in the water remediation process is significant due to higher reaction rates, a larger surface area for reactions, a tendency to adsorb contaminants of various types, and selectivity for aromatic compounds. Several forms of carbon nanomaterials are explained, such as carbon nanotubes, beads, fibers, and nanoporous carbon.

Carbon Nanotubes (CNTs), out of the various carbon nanomaterials, received much attention in research in the past. CNTs are graphene sheets in the form of cylinders with a 1 nm diameter. CNTs show remarkable adsorption capacities owing to a large surface area with porous structures. CNTs are reported to be involved in the adsorption of various organic, inorganic substances and metal cations present in water, for example, dichlorobenzene [49], ethylbenzene, Zn(II) [50], Pb(II), Cu(II), Cd(II) [51], and dyes. Two types of CNTs have been identified so far. One consists of multi-walled carbon nanotubes (MWCNTs), comprising of many concentric layers as a cylinder and a space of 0.34 nm between the two concentric layers. The other consists of single-walled carbon nanotubes (SWCNTs), made up of single-layered graphene sheets rolled up as a cylinder [52]. Although CNTs cover a wide range of contaminants present in water, certain limitations also confine them because CNTs need a support, matrix, or base to enable their action. CNTs can be combined with other metals or anything to improve their reactivity. This functionalization increases

Introduction: Nanomaterials and Their Role in Water Treatment

the number of active sites on the surfaces of the CNTs, and dispersibility is found to be enhanced [53]. In the literature, the combination of magnetic iron oxide with CNT support has been reported by Gupta et al. [54]. When the magnetic and adsorption tendencies are combined, the "composite" adsorbent can readily remove chromium from water. The magnetic material can be easily recovered after degradation when subjected to an external magnetic field. Its production is quite limited, and the cost is high. This feature also does not attract much usage.

Graphene is a carbonaceous material with a planar sheet-like structure because of the sp^2 hybridization of carbon atoms constituting graphene. It is one of the best membranes for filtration because of its atomic thickness. O'Hern and associates [55] have fabricated a monolayer graphene composite membrane with a high filtration area. The graphene layer possesses a pore size of around 1-15 nm, allowing the permeability of selected materials. Graphene can be used as graphene oxide, reduced graphene oxide, and pristine graphene. Graphene can also effectively adsorb dyes, heavy metals, surfactants, and pharmaceuticals from water. Owing to the high surface-to-weight ratio coupled with chemical stability, graphene behaves as a remarkable agent for the adsorption of contaminants, as mentioned earlier. GO, and rGO have been found to exhibit remarkable adsorption tendency against non-toxic surfactants [56]. Cysteamine functionalized rGO has been successfully tested to remove Hg^{2+} from water [57]. Lv and co-workers prepared modified GO and tested it against several dyes, ciprofloxacin, and Cu(II) ions in water [58]. Nanozirconium carbide was prepared and has been found to show significant adsorption against pharmaceuticals [59].

Along with adsorption, graphene-based nanomaterials have also found application in membrane filtration. Graphene/GO/rGO-based nanocomposites are also tested for the photocatalytic degradation of organic contaminants. Functionalization of GO nanosheets with AgNPs exhibited strong antimicrobial behavior against *E. coli* and *S. aureus* [60]. Monolayer graphene sheets cannot withstand the pressure condition for a longer period. To overcome this challenge, multi-layered GO can be produced. The release of graphene into the water can also cause toxicity problems, so there is still a need to fabricate such graphene sheets that can be successful against water contamination and are environmentally friendly.

1.3.4 Nanocomposites

Although all the nanomaterials mentioned above are widely used for the water treatment process, these are also associated with certain drawbacks such as iron NPs showing a tendency of accumulation, oxidation, and the tedious process after the treatment. Metal oxide NPs like TiO_2 and ZnO absorb light only in the UV region due to high bandgap energy. CNTs need support for their activity, and additionally, their costs are very high.

For the construction of nanocomposites, two types of nanomaterials are combined, one is nano Fe particles, and another one is CNT. The resultant nanocomposites bear the characteristics of both these particles. Because of the CNTs, adsorption tendency reaches a higher level, and it was reported in the literature that it could adsorb NO^{3-} ions from the wastewater [61]. The magnetic property shown by Fe nanomaterials makes them separable from the treated water using an external magnetic field. There is a need to develop a desirable nanocomposite that is non-toxic, stable, cheap, and that serves the purpose of water decontamination well.

1.4 WATER TREATMENT METHODOLOGIES

So far, various types of NPs involved in water remediation have been discussed. Here, we discuss the methodologies used to show the action of NPs against water pollutants. This includes nanophotocatalysis, nanosorbents, and nanomembranes.

1.4.1 NANOPHOTOCATALYSTS

As NPs provide a large surface area for any reaction to occur, these can be proven to show better catalytic activities. For wastewater treatment, in the presence of light, nanophotocatalysts play a vital role in the degradation of compounds. Photocatalysts stimulate any chemical transformation in the presence of light. Nanophotocatalysts effectively produce oxidizing species which actively participate in the degradation of contaminants in the water. Nanomaterials are very effective for removing pollutants such as dyes, chlorpyrifos, nitroaromatics, and the like. SiO_2, ZnO_2, and TiO_2 are examples of some commonly reported nanometallic oxides which fulfil the function of nanophotocatalysis. A literature survey revealed TiO_2 is one of the best photocatalysts out of these metallic oxides due to its abundancy, stability, low price, and non-toxic tendency. Naturally, TiO_2 is found in three states: anatase, rutile, and brookite. Anatase has been reported to be a better nanophotocatayst than the other two. The bandgap of the anatase state is 3.2 eV which corresponds to the UV region. It has been associated with removing several categories of pollutants from water, for example, azo dyes, aromatic compounds containing phenol functionality, chlorinated aromatics, and the like.

Besides TiO_2, ZnO has also been recognized as an effective photocatalyst and successfully examined against water contaminants. Composites made up of CdS/TiO_2 in the presence of visible light have been effectively tested against DMSO [62]. Nano ZnO combined with Pd has also shown remarkable photocatalytic activity against *E. coli* in water [63]. Certain modifications can be made to increase the photocatalytic activity of metal oxide particles in the visible region, such as incorporating metal/metal ions and carbonaceous-based materials, and so forth.

As a result of the degradation of organic pollutants by TiO_2, CO_2 and inorganic ions are formed. In the presence of UV light, titanium dioxide absorbs wavelengths corresponding to its bandgap, and electron-hole (e^--h^+) pairs are formed. Holes are produced in the valence band, whereas electrons move in the conduction band owing to electronic excitation.

$$TiO_2 + h\nu \rightarrow e^-_{cb} + h^+_{vb}$$

$$O_2 + e^-_{cb} \rightarrow O_2^-$$

$$H_2O + h^+_{vb} \rightarrow \cdot OH + H^+$$

When charge separation is continued, the e^--h^+ move towards the catalyst's surface, where h^+ reacts with water to generate $HO\cdot$ and e^- is absorbed by O_2 to produce O^{2-} (Figure 1.4).

Both hydroxyl free radicals and superoxide ions take part in the degradation of organic contaminants. Sometimes, organic pollutants are not degraded completely due to the formation of stable intermediate products. To facilitate complete degradation, TiO_2 particles are irradiated for a longer period. TiO_2 has better photostability than other nanophotocatalysts such ZnO, metal sulfide, Cu-based materials, and so on. These photocatalysts participate in the process of photocorrosion. Upon irradiation, these materials may undergo reduction or oxidation, and a change in oxidation state occurs by generating h^+ or e^-, leading to the decomposition of photocatalysts. To solve this problem, nanophotocatalysts should be changed into the form of nanocomposites to get photochemically stable products. Researchers are also focusing on developing such photocatalysts that are neither toxic in water nor produce toxic products after degradation, along with the materials that can absorb in the visible region. For this reason, heterojunctions were developed to improve the light-harvesting properties. The composite α-Fe_2O_3/g-C_3N_4 was synthesized, which showed good degradation efficiency due to enhanced charge transferability (Figure 1.5) [64].

Nanophotocatalysts work at 25°C and detoxify water effectively, provide a large surface area for reaction, are less toxic, are low cost, are chemically stable,

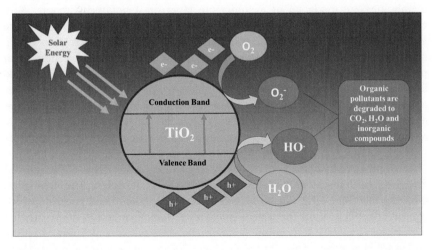

FIGURE 1.4 Role of TiO_2 as nanophotocatalyst for degradation of organic pollutants.

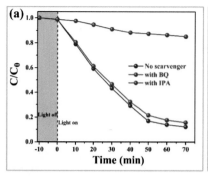 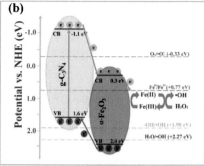

FIGURE 1.5 (a) Degradation of phenol over α-Fe$_2$O$_3$/g-C$_3$N$_4$ in the presence of radical scavengers. (b) The heterojunction mechanism of α-Fe$_2$O$_3$/g-C$_3$N$_4$ (proposed) ([64]). License: http://creativecommons.org/licenses/by/4.0/).

and show high rates of catalytic reactions. There are still certain drawbacks with current practice such as non-recovery of the catalyst after degradation, and quantum size effect which enhances the bandgap energy. To recover the catalysts from the degraded system, magnetic nanophotocatalysts can be used.

1.4.2 Nanomembranes

These membranes are fabricated with various nanofibers to remove particles present in water. They are also used for the process of reverse osmosis. Various types of nanomaterials can be impregnated with various polymer supports to produce a variety of nanomembranes. These membranes are pervious to water and perform the function of nanofiltration, reverse osmosis, and so forth. The membranes are made up of a composite layer on a polymer support. As a composite layer, carbon nanotubes are used. CNTs are embedded within the polymer matrix. This fabrication makes up the membrane formation which shows significant fouling resistance. CNTs show remarkable antimicrobial properties and reduce fouling. On polymer membranes, some dopants may also be added. These might be nanomaterials like zeolite, Al$_2$O$_3$, TiO$_2$, and the like, resulting in the formation of a modified membrane with fouling resistance and a hydrophilic surface [65]. The doping of AgNPs with a polymer produces the membrane, reduces bacterial film formation on the membrane and inactivates viruses.

Over time, the nanomembrane methodology faces the problem of fouling and clogging, which the addition of super hydrophilic NPs can address. The membranes made up of Al$_2$O$_3$, TiO$_2$, Ag, and CNTs have hydrophilicity and porosity, which can be effectively used for water remediation. Nanomembranes are associated with certain qualities: uniformity, homogeneity, ease of use, quickness of action, and order in reaction. These properties have drawn attention to their use. These

Introduction: Nanomaterials and Their Role in Water Treatment 15

membranes can also be incorporated with nanophotocatalysts, such as TiO_2, which can degrade organic pollutants and disinfect the water.

Membranes can also be combined with several catalytic materials to achieve multifunctional activities. Introducing the catalytic membrane can improve the membrane's selectivity, decomposition rate, and fouling resistance. To this effect, a composite membrane can be constructed having Fe-catalyzed based free radicals within the pore membrane. It can be used to serve both the purpose of the detoxification and degradation of organic pollutants. Immobilizing various nano-metallic particles on membranes made up of chitosan, cellulose acetate, and the like, was found to be effective in the degradation of toxic compounds in water because of properties such as high reactivity, lack of particle accumulation, and surface reduction [66]. Nanocomposite membranes comprised of palladium acetate and polyetherimide were successfully tested for water decontamination [67]. Various nanomembranes such as nanofiber, nanocomposite, aqua-porin-based, self-assembling, and nanofiltration types have been identified and used for ultrafiltration and prefiltration color, odor, and hardness removal, and so on [68,69]. Nanofiber membranes are highly porous, bactericidal, and permeable but also possess the property of pore-blocking and the release of nanofibers in the water. Membranes made up of nanocomposites are permeable, hydrophilic, and have high-grade fouling resistance with excellent thermal and mechanical stability. Nevertheless, the use of oxidizing nanomaterial may cause the discharge of NPs. Membranes based on aqua-porin show ionic selectivity and permeability, but these are not strong mechanically. Self-assembling membranes have homogeneous porosity, but their availability is much less. Nanofiltration membranes show good selectivity and can be used to remove the color, odor, and hardness of water but are also associated with a tendency towards membrane blocking. The use of nanomembranes is a better method for removing hardness than conventional methods, as these do not require Na^+ ions to remove $Ca(II)$ or $Mg(II)$ from water.

One of the major drawbacks of nanomembranes is the fouling of the membrane after some time, making its use costly and less efficient. Another drawback is the instability of the membrane after a period, so there is a need to change the membrane to carry out water filtration continuously. This causes an increase in the cost of the procedure and in the chances of contamination. More research is required to generate nanomembranes, which are more selective and resistive, to prevent fouling. Using zwitterions on the surface can overcome some of the drawbacks. So, there is a continuous need to fabricate multifunctional membranes that are lower in cost and much more efficient even at industrial scales.

1.4.3 Nanosorbents

Sorbents are substances that have sorption capacity and are more likely to be used in the water remediation process. Many nanomaterials are found to be actively endowed in the sorption phenomena. Carbon-based nanomaterials such as C-black and graphenes are considered to be the best nanosorbents amongst all of the reported substances [70]. Furthermore, zero-valent metals, metallic oxides, and various composites made up of two different materials are also useful in this field [71]. Carbon nanotubes- SWCNTs and MWCNTs have multiple active sites on their surfaces. Pollutants become adhered to on their surfaces and are thus degraded. These multiple active sites also do not allow the aggregation of contaminants in water. CNTs are hydrophobic at their surface, so they are effective against contaminants for their adsorption. For removing inorganic and organic substances from water, polymer-based nanoadsorbents such as dendrimers are also significant, for example, the reduction of copper ions by the dendrimer system has been reported in the literature [72]. Dendrimers are found to exhibit almost one hundred percent efficiency against dyes or organic pollutants.

Amongst nanosorbents, zeolites also hold a good position for removing various ions and microbes from water. Additionally, magnetic nanosorbents can be synthesized by coating magnetic NPs on certain ligands and find their application in removing organic pollutants [73]. Many nanosorbents have been identified in the literature, such as carbon-based, graphite-oxide, polymeric nano-sorbents, nanometal oxides, and fibers in treating Ni(II), dyes organic pollutants, heavy metals arsenic, and other heavy metals, respectively [74]. Various methods can be adapted to regenerate these exhausted nanosorbents, for example, ion exchange, magnetic effects, and the like.

Using carbon-based materials has some health hazards, which depend upon the structure, chemical stability, and surface structure of the nanosorbents. GOs are the best alternative for water pollutant removal. Nanosorbents show significant physicochemical properties in water for pollutant degradation and provide a large surface area for the degradation or adsorption of pollutants. These particles show a high rate of reaction. Degradation occurs in a concise period. However, for the commercial use of nanosorbents certain challenges need to be answered, such as the selectivity of sorbent materials, their stability, and the amount of time taken by them to operate. There is an immediate requirement for the fabrication of nanoadsorbents including new emerging materials such as MOFs that can be used effectively to remove toxic ions and compounds from water [14].

1.5 CONCLUSION

Nanomaterials hold a good position as an important solution to the emerging problem of potable drinking water, which has been increasing day by day because of massive population growth, depletion in natural freshwater sources, and the increasing level of pollutants in water. Nanomaterials as particles, catalysts, sorbents, and membranes serve the purpose of water remediation including several processes such as the removal of organic, inorganic, and toxic metals from water, its decontamination from pathogenic and non-pathogenic microbes, and so forth. Due to nano size and large surface area, their chemical reactivity rate on pollutant degradation is many times higher than any conventional source.

So far, many types of nanomaterials have been used for the decontamination process. But it has been inferred from several research papers that carbon-based nanomaterials such as graphenes and carbon nanotubes occupy a good position for utilization potential as dynamic agents against water pollutants such as some surfactant antibiotics. These can also remove heavy metal ions, which are present in unacceptable concentrations in water. The insertion of various functional groups on carbon nanotubes also enables them to degrade different pollutants. Additionally, silver, iron, and zinc-based nanomaterials, including metals and metal oxides, activate in light and effectively degrade pollutants, especially commercial dyes.

Although many NPs have been identified so far, there is still a requirement to search for new nanomaterials that need to be effective, cheap, environment friendly, easy to use, and easy to store. In the future, nanomaterials might supersede conventional methods because of the potential of the NPs to be utilize as the best alternative for water remediation.

REFERENCES

[1] Buzea, Cristina, Ivan I. Pacheco, and Kevin Robbie. "Nanomaterials and Nanoparticles: Sources and Toxicity." *Biointerphases*, 2, no. 4 (2007), MR17–71. https://doi.org/10.1116/1.2815690.

[2] Kabir, Ehsanul, Vanish Kumar, Ki Hyun Kim, Alex C.K. Yip, and J. R. Sohn. "Environmental Impacts of Nanomaterials." *Journal of Environmental Management*, 225 (2018), 261–271. https://doi.org/10.1016/j.jenvman.2018.07.087.

[3] Joshi, Priyanka, Meena Nemiwal, Abdullah A. Al-Kahtani, Mohd Ubaidullah, and Dinesh Kumar. "Biogenic AgNPs for the Non-Cross-Linking Detection of Aluminum in Aqueous Systems." *Journal of King Saud University—Science* 33, no. 6 (2021), 101527. https://doi.org/10.1016/j.jksus.2021.101527.

[4] Nemiwal, Meena, and Dinesh Kumar. "Metal Organic Frameworks as Water Harvester from Air: Hydrolytic Stability and Adsorption Isotherms." *Inorganic Chemistry Communications*, 122 (2020), 108279. https://doi.org/10.1016/j.inoche.2020.108279.

[5] Nemiwal, Meena, Tian C. Zhang, and Dinesh Kumar. "Recent Progress in g-C_3N_4, TiO_2 and ZnO Based Photocatalysts for Dye Degradation: Strategies to Improve Photocatalytic Activity." *Sci. Total Environ.*, 767 (2021), 144896. https://doi.org/10.1016/j.scitotenv.2020.144896.

[6] Kumari, Sandhya, Kritika S. Sharma, Meena Nemiwal, Suphiya Khan, and Dinesh Kumar. "Simultaneous Detection of Aqueous Aluminum(III) and Chromium(III) Using Persea Americana Reduced and Capped Silver Nanoparticles." *International Journal of Phytoremediation* (2021), 24(8): 1–14. https://doi.org/10.1080/15226514.2021.1977911.

[7] Nemiwal, Meena, and Dinesh Kumar. "Recent Progress on Electrochemical Sensing Strategies as Comprehensive Point-Care Method." *Monatshefte Fur Chemie*, 152, no. 1 (2021), 118. https://doi.org/10.1080/15226514.2021.1977911.

[8] Ahn, Youngmee, and Yonghoon Jun. "Measurement of Pain-like Response to Various NICU Stimulants for High-Risk Infants." *Early Human Development* 83, no. 4 (2007), 255–62. https://doi.org/10.1016/j.earlhumdev.2006.05.022.

[9] Nemiwal, Meena, Tian C. Zhang, and Dinesh Kumar. "Pectin Modified Metal Nanoparticles and Their Application in Property Modification of Biosensors." *Carbohydrate Polymer Technologies and Applications*, 2 (2021), 100164. https://doi.org/10.1016/j.carpta.2021.100164.

[10] Jindal, Himani, Dinesh Kumar, Mika Sillanpaa, and Meena Nemiwal. "Current Progress in Polymeric Graphitic Carbon Nitride-Based Photocatalysts for Dye Degradation." *Inorganic Chemistry Communications*, 131 (2021), 108786. https://doi.org/10.1016/j.inoche.2021.108786.

[11] Nemiwal, Meena, Vijayalakshmi Gosu, Tian C. Zhang, and Dinesh Kumar. "Metal Organic Frameworks as Electrocatalysts: Hydrogen Evolution Reactions and Overall Water Splitting." *International Journal of Hydrogen Energy*, 46 (2021), 10216–38. https://doi.org/10.1016/j.ijhydene.2020.12.146.

[12] Nemiwal, Meena, and Dinesh Kumar. "Metal Organic Frameworks as Water Harvester from Air: Hydrolytic Stability and Adsorption Isotherms." *Inorganic Chemistry Communications*, 122 (2020), 108279. https://doi.org/10.1016/j.inoche.2020.108279.

[13] Nemiwal, Meena, Vijayalakshmi Gosu, Ankita Dhillon, and Dinesh Kumar. "Environmental Applications of Metal-Organic Frameworks: Recent Advances and Challenges." In *ACS Symposium Series*, 1394 (2021), 299–318. https://doi.org/10.1021/bk-2021-1394.ch012.

[14] Singh, Niharika, Ankita Dhillon, Meena Nemiwal, and Dinesh Kumar. "Metal-Organic Frameworks for Water Decontamination and Reuse: A Dig at Heavy Metal Ions and Organic Toxins." In *ACS Symposium Series*, 1395 (2021), 77–124. https://doi.org/10.1021/bk-2021-1395.ch004.

[15] Nemiwal, Meena, Venu Sharma, and Dinesh Kumar. "Improved Designs of Multifunctional Covalent-Organic Frameworks: Hydrogen Storage, Methane Storage, and Water Harvesting." *Mini-Reviews in Organic Chemistry*, 18, no. 8 (2020), 1026–36. https://doi.org/10.2174/1570193x17999201127105752.

[16] Ahmad, Akil, Siti Hamidah Mohd-Setapar, Chuo Sing Chuong, Asma Khatoon, Waseem A. Wani, Rajeev Kumar, and Mohd Rafatullah. "Recent Advances in New Generation Dye Removal Technologies: Novel Search for Approaches to Reprocess Wastewater." *RSC Advances*, 5 (2015), 30801–18. https://doi.org/10.1039/c4ra16959j.

[17] Daer, Sahar, Jehad Kharraz, Adewale Giwa, and Shadi Wajih Hasan. "Recent Applications of Nanomaterials in Water Desalination: A Critical Review and Future Opportunities." *Desalination*, 367 (2015), 37–48. https://doi.org/10.1016/j.desal.2015.03.030.

[18] Chen, Cheng, Yuqi Fan, Jianhang Gu, Liming Wu, Stefano Passerini, and Liqiang Mai. "One-Dimensional Nanomaterials for Energy Storage." *Journal of Physics D: Applied Physics*, 51 (2018). https://doi.org/10.1088/1361-6463/aaa98d.

[19] Chatterjee, A., and B. L. Deopura. "Carbon Nanotubes and Nanofibre: An Overview." *Fibers and Polymers*, 3 (2002), 134–39. https://doi.org/10.1007/BF02912657.

[20] Li, Xiao, Yuqing Li, Qiu, Qirui Wen, Qi Zhang, Wenjing Yang, Lihui Yuwen, Lixing Weng, and Lianhui Wang. "Efficient Biofunctionalization of MoS_2 Nanosheets with Peptides as Intracellular Fluorescent Biosensor for Sensitive Detection of Caspase-3 Activity." *Journal of Colloid and Interface Science*, 543 (2019), 96–105. https://doi.org/10.1016/j.jcis.2019.02.011.

[21] Gao, Xin, Hongyan Yue, Shanshan Song, Shuo Huang, Bing Li, Xuanyu Lin, Erjun Guo, et al. "3-Dimensional Hollow Graphene Balls for Voltammetric Sensing of Levodopa in the Presence of Uric Acid." *Microchimica Acta*, 185, no. 2 (2018), 1-8. https://doi.org/10.1007/s00604-017-2644-y.

[22] Kim, Dong Wook, Thai Minh Duy Le, Sang Moon Lee, Hae Jin Kim, Yoon Joo Ko, Ji Hoon Jeong, Thavasyappan Thambi, Doo Sung Lee, and Seung Uk Son. "Microporous Organic Nanoparticles Anchoring CeO_2 Materials: Reduced Toxicity and Efficient Reactive Oxygen Species-Scavenging for Regenerative Wound Healing." *ChemNanoMat*, 6, no. 7 (2020), 1104–10. https://doi.org/10.1002/cnma.202000067.

[23] Kesse, Samuel, Kofi Oti Boakye-Yiadom, Belynda Owoya Ochete, Yaw Opoku-Damoah, Fahad Akhtar, Mensura Sied Filli, Muhammad Asim Farooq, et al. "Mesoporous Silica Nanomaterials: Versatile Nanocarriers for Cancer Theranostics and Drug and Gene Delivery." *Pharmaceutics*, 11, no. 2 (2019), 77. https://doi.org/10.3390/pharmaceutics11020077.

[24] Sanati, Alireza, Keyvan Raeissi, and Fathallah Karimzadeh. "A Cost-Effective and Green-Reduced Graphene Oxide/Polyurethane Foam Electrode for Electrochemical Applications." *FlatChem*, 20 (2020), 100162. https://doi.org/10.1016/j.flatc.2020.100162.

[25] Heijden, A. E.D.M. van der. "Developments and Challenges in the Manufacturing, Characterization and Scale-up of Energetic Nanomaterials—A Review." *Chemical Engineering Journal*, 350 (2018), 939–48. https://doi.org/10.1016/j.cej.2018.06.051.

[26] Sondi, Ivan, and Branka Salopek-Sondi. "Silver Nanoparticles as Antimicrobial Agent: A Case Study on E. Coli as a Model for Gram-Negative Bacteria." *Journal of Colloid and Interface Science*, 275, no. 1 (2004), 177–82. https://doi.org/10.1016/j.jcis.2004.02.012.

[27] Danilczuk, M., A. Lund, J. Sadlo, H. Yamada, and J. Michalik. "Conduction Electron Spin Resonance of Small Silver Particles." *Spectrochimica Acta—Part A: Molecular and Biomolecular Spectroscopy*, 63, no. 1 (2006), 189–91. https://doi.org/10.1016/j.saa.2005.05.002.

[28] Dhanalekshmi, K. I., and K. S. Meena. "DNA Intercalation Studies and Antimicrobial Activity of $Ag@ZrO_2$ Core-Shell Nanoparticles in Vitro." *Materials Science and Engineering C*, 59 (2016), 1063–68. https://doi.org/10.1016/j.msec.2015.11.027.

[29] Ren, Dianjun, and James A. Smith. "Retention and Transport of Silver Nanoparticles in a Ceramic Porous Medium Used for Point-of-Use Water Treatment." *Environmental Science and Technology*, 47, no. 8 (2013), 3825–32. https://doi.org/10.1021/es4000752.

[30] Ferreira, Aline Marques, Érica Barbosa Roque, Fabiana Valéria da Fonseca, and Cristiano Piacsek Borges. "High Flux Microfiltration Membranes with Silver Nanoparticles for Water Disinfection." *Desalination and Water Treatment*, 56, no. 13 (2015), 3590–98. https://doi.org/10.1080/19443994.2014.1000977.

[31] Wang, Yu, Zhanqiang Fang, Yuan Kang, and Eric Pokeung Tsang. "Immobilization and Phytotoxicity of Chromium in Contaminated Soil Remediated by CMC-Stabilized NZVI." *Journal of Hazardous Materials*, 275 (2014), 230–37. https://doi.org/10.1016/j.jhazmat.2014.04.056.

[32] Zhao, Yan Li, and J. Fraser Stoddart. "Noncovalent Functionalization of Single-Walled Carbon Nanotubes." *Accounts of Chemical Research*, 42, no. 8 (2009), 1161–71. https://doi.org/10.1021/ar900056z.

[33] Liou, Ya Hsuan, Shang Lien Lo, Chin Jung Lin, Wen Hui Kuan, and Shih Chi Weng. "Chemical Reduction of an Unbuffered Nitrate Solution Using Catalyzed and Uncatalyzed Nanoscale Iron Particles." *Journal of Hazardous Materials*, 127, no. 1–3 (2005), 102–10. https://doi.org/10.1016/j.jhazmat.2005.06.029.

[34] Bokare, Varima, Ju lim Jung, Yoon Young Chang, and Yoon Seok Chang. "Reductive Dechlorination of Octachlorodibenzo-p-Dioxin by Nanosized Zero-Valent Zinc: Modeling of Rate Kinetics and Congener Profile." *Journal of Hazardous Materials*, 250–251 (2013), 397–402. https://doi.org/10.1016/j.jhazmat.2013.02.020.

[35] Fujishima, Akira, and Kenichi Honda. "Electrochemical Photolysis of Water at a Semiconductor Electrode." *Nature*, 238, no. 5358 (1972), 37–38. https://doi.org/10.1038/238037a0.

[36] Gar Alalm, Mohamed, Ahmed Tawfik, and Shinichi Ookawara. "Comparison of Solar TiO_2 Photocatalysis and Solar Photo-Fenton for Treatment of Pesticides Industry Wastewater: Operational Conditions, Kinetics, and Costs." *Journal of Water Process Engineering*, 8 (2015), 55–63. https://doi.org/10.1016/j.jwpe.2015.09.007.

[37] Chen, Zengping, Yaru Li, Meng Guo, Fengyun Xu, Peng Wang, Yu Du, and Ping Na. "One-Pot Synthesis of Mn-Doped TiO_2 Grown on Graphene and the Mechanism for Removal of Cr(VI) and Cr(III)." *Journal of Hazardous Materials*, 310 (2016), 188–98. https://doi.org/10.1016/j.jhazmat.2016.02.034.

[38] Foster, Howard A., Iram B. Ditta, Sajnu Varghese, and Alex Steele. "Photocatalytic Disinfection Using Titanium Dioxide: Spectrum and Mechanism of Antimicrobial Activity." *Applied Microbiology and Biotechnology*, 90 (2011), 1847–68. https://doi.org/10.1007/s00253-011-3213-7.

[39] Seery, Michael K., Reenamole George, Patrick Floris, and Suresh C. Pillai. "Silver Doped Titanium Dioxide Nanomaterials for Enhanced Visible Light Photocatalysis." *Journal of Photochemistry and Photobiology A: Chemistry*, 189, no. 2–3 (2007), 258–63. https://doi.org/10.1016/j.jphotochem.2007.02.010.

[40] Liu, Chin Chuan, Yung Hsu Hsieh, Pao Fan Lai, Chia Hsin Li, and Chao Lang Kao. "Photodegradation Treatment of Azo Dye Wastewater by UV/TiO_2 Process." *Dyes and Pigments*, 68, no. 2–3 (2006), 191–95. https://doi.org/10.1016/j.dyepig.2004.12.002.

[41] Wang, Qianqian, Xiaoting Wang, Zhaohui Wang, Jun Huang, and Yong Wang. "PVDF Membranes with Simultaneously Enhanced Permeability and Selectivity by Breaking the Tradeoff Effect via Atomic Layer Deposition of TiO_2." *Journal of Membrane Science*, 442 (2013), 57–64. https://doi.org/10.1016/j.memsci.2013.04.026.

[42] Janotti, Anderson, and Chris G. van De Walle. "Fundamentals of Zinc Oxide as a Semiconductor." *Reports on Progress in Physics*, 72, no. 12 (2009), 126501. https://doi.org/10.1088/0034-4885/72/12/126501.

[43] Lee, Kian Mun, Chin Wei Lai, Koh Sing Ngai, and Joon Ching Juan. "Recent Developments of Zinc Oxide Based Photocatalyst in Water Treatment Technology: A Review." *Water Research*, 88 (2016), 428–48. https://doi.org/10.1016/j.watres.2015.09.045.

[44] Kumar, Parveen, Vijesh Tomar, Dinesh Kumar, Raj Kumar Joshi, and Meena Nemiwal. "Magnetically Active Iron Oxide Nanoparticles for Catalysis of Organic Transformations: A Review." *Tetrahedron*, 106–107 (January 29, 2022), 132641. https://doi.org/10.1016/j.tet.2022.132641.
[45] Warner, Cynthia L., R. Shane Addleman, Anthony D. Cinson, Timothy C. Droubay, Mark H. Engelhard, Michael A. Nash, Wassana Yantasee, and Marvin G. Warner. "High-Performance, Superparamagnetic, Nanoparticle-Based Heavy Metal Sorbents for Removal of Contaminants from Natural Waters." *ChemSusChem*, 3, no. 6 (2010), 749–57. https://doi.org/10.1002/cssc.201000027.
[46] Ge, Fei, Meng Li, Hui Ye, and Bao Xiang Zhao. "Effective Removal of Heavy Metal Ions Cd^{2+}, Zn^{2+}, Pb^{2+}, Cu^{2+} from Aqueous Solution by Polymer-Modified Magnetic Nanoparticles." *Journal of Hazardous Materials*, 211–212 (2012), 366–72. https://doi.org/10.1016/j.jhazmat.2011.12.013.
[47] Liang, Hanfeng, Binbin Xu, and Zhoucheng Wang. "Self-Assembled 3D Flower-like α-Fe_2O_3 Microstructures and Their Superior Capability for Heavy Metal Ion Removal." *Materials Chemistry and Physics*, 141, no. 2–3 (2013), 727–34. https://doi.org/10.1016/j.matchemphys.2013.05.070.
[48] Shahmirzaee, Mozhgan, Abdolhossein Hemmati-Sarapardeh, Maen M. Husein, Mahin Schaffie, and Mohammad Ranjbar. "Magnetic γ-Fe_2O_3/ZIF-7 Composite Particles and Their Application for Oily Water Treatment." *ACS Omega*, 7, no. 4 (February 1, 2022), 3700–12. https://doi.org/10.1021/acsomega.1c06382.
[49] Peng, Xianjia, Yanhui Li, Zhaokun Luan, Zechao Di, Hongyu Wang, Binghui Tian, and Zhiping Jia. "Adsorption of 1,2-Dichlorobenzene from Water to Carbon Nanotubes." *Chemical Physics Letters*, 376, no. 1–2 (2003), 154–58. https://doi.org/10.1016/S0009-2614(03)00960-6.
[50] Cho, Hyun Hee, Kevin Wepasnick, Billy A. Smith, Fazlullah K. Bangash, D. Howard Fairbrother, and William P. Ball. "Sorption of Aqueous Zn[II] and Cd[II] by Multiwall Carbon Nanotubes: The Relative Roles of Oxygen-Containing Functional Groups and Graphenic Carbon." *Langmuir*, 26, no. 2 (2010), 967–81. https://doi.org/10.1021/la902440u.
[51] Li, Yan Hui, Jun Ding, Zhaokun Luan, Zechao Di, Yuefeng Zhu, Cailu Xu, Dehai Wu, and Bingqing Wei. "Competitive Adsorption of Pb^{2+}, Cu^{2+} and Cd^{2+} Ions from Aqueous Solutions by Multiwalled Carbon Nanotubes." In *Carbon*, 41 (2003), 2787–92. https://doi.org/10.1016/S0008-6223(03)00392-0.
[52] Zhao, Yan Li, and J. Fraser Stoddart. "Noncovalent Functionalization of Single-Walled Carbon Nanotubes." *Accounts of Chemical Research*, 42, no. 8 (2009), 1161–71. https://doi.org/10.1021/ar900056z.
[53] Adeleye, Adeyemi S., Jon R. Conway, Kendra Garner, Yuxiong Huang, Yiming Su, and Arturo A. Keller. "Engineered Nanomaterials for Water Treatment and Remediation: Costs, Benefits, and Applicability." *Chemical Engineering Journal*, 286 (2016), 640–62. https://doi.org/10.1016/j.cej.2015.10.105.
[54] Gupta, V. K., Shilpi Agarwal, and Tawfik A. Saleh. "Chromium Removal by Combining the Magnetic Properties of Iron Oxide with Adsorption Properties of Carbon Nanotubes." *Water Research*, 45, no. 6 (2011), 2207–12. https://doi.org/10.1016/j.watres.2011.01.012.
[55] O'Hern, Sean C., Cameron A. Stewart, Michael S. H. Boutilier, Juan Carlos Idrobo, Sreekar Bhaviripudi, Sarit K. Das, Jing Kong, Tahar Laoui, Muataz Atieh, and Rohit Karnik. "Selective Molecular Transport through Intrinsic Defects in a

Single Layer of CVD Graphene." *ACS Nano*, 6, no. 11 (2012), 10130–38. https://doi.org/10.1021/nn303869m.

[56] Prediger, Patricia, Thais Cheminski, Tauany de Figueiredo Neves, William Bardelin Nunes, Livia Sabino, Carolina Siqueira Franco Picone, Rafael L. Oliveira, and Carlos Roque Duarte Correia. "Graphene Oxide Nanomaterials for the Removal of Non-Ionic Surfactant from Water." *Journal of Environmental Chemical Engineering*, 6, no. 1 (2018), 1536–45. https://doi.org/10.1016/j.jece.2018.01.072.

[57] Yap, Pei Lay, Shervin Kabiri, Diana N.H. Tran, and Dusan Losic. "Selective and Highly Efficient Adsorption of Mercury." *ACS Applied Materials and Interfaces* (2019). https://doi.org/10.1021/acsami.8b17131.

[58] Lv, Mingqian, Liwei Yan, Cheng Liu, Chunjiao Su, Qilin Zhou, Xiao Zhang, Yi Lan, et al. "Non-Covalent Functionalized Graphene Oxide (GO) Adsorbent with an Organic Gelator for Co-Adsorption of Dye, Endocrine-Disruptor, Pharmaceutical and Metal Ion." *Chemical Engineering Journal*, 349 (2018), 791–99. https://doi.org/10.1016/j.cej.2018.04.153.

[59] Zhang, Bingjie, Jiawen Ji, Xue Liu, Changsheng Li, Meng Yuan, Jingyang Yu, and Yongqiang Ma. "Rapid Adsorption and Enhanced Removal of Emodin and Physcion by Nano Zirconium Carbide." *Science of the Total Environment*, 647 (2019), 57–65. https://doi.org/10.1016/j.scitotenv.2018.07.422.

[60] Bao, Qi, Dun Zhang, and Peng Qi. "Synthesis and Characterization of Silver Nanoparticle and Graphene Oxide Nanosheet Composites as a Bactericidal Agent for Water Disinfection." *Journal of Colloid and Interface Science*, 360, no. 2 (2011), 463–70. https://doi.org/10.1016/j.jcis.2011.05.009.

[61] Azari, Ali, Ali Akbar Babaei, Roshanak Rezaei-Kalantary, Ali Esrafili, Mojtaba Moazzen, and Babak Kakavandi. "Nitrate Removal from Aqueous Solution Using Carbon Nanotubes Magnetized by Nano Zero-Valent Iron." *Journal of Mazandaran University of Medical Sciences*, 23, no. 2 (2014), 14–27.

[62] Li, Xin, Ting Xia, Changhui Xu, James Murowchick, and Xiaobo Chen. "Synthesis and Photoactivity of Nanostructured CdS-TiO_2 Composite Catalysts." *Catalysis Today*, 225 (2014), 64–73. https://doi.org/10.1016/j.cattod.2013.10.086.

[63] Berekaa, Mahmoud M. "Nanotechnology in Wastewater Treatment; Influence of Nanomaterials on Microbial Systems." *International Journal of Current Microbiology and Applied Sciences*, 5, no. 1 (2016), 713–26. https://doi.org/10.20546/ijcmas.2016.501.072.

[64] Ge, Fuxiang, Xuehua Li, Mian Wu, Hui Ding, and Xiaobing Li. " A Type II Heterojunction α-Fe_2O_3/g-C_3N_4 for the Heterogeneous Photo-Fenton Degradation of Phenol." *RSC Advances*, 12, no. 14 (2022), 8300–09. https://doi.org/10.1039/d1ra09282k.

[65] Waduge, Pradeep, Joseph Larkin, Moneesh Upmanyu, Swastik Kar, and Meni Wanunu. "Programmed Synthesis of Freestanding Graphene Nanomembrane Arrays." *Small*, 11, no. 5 (2015), 597–603. https://doi.org/10.1002/smll.201402230.

[66] Jawed, Aquib, Varun Saxena, and Lalit M. Pandey. "Engineered Nanomaterials and Their Surface Functionalization for the Removal of Heavy Metals: A Review." *Journal of Water Process Engineering*, 33 (2020), 101009. https://doi.org/10.1016/j.jwpe.2019.101009.

[67] Umar, Khalid, Azmi Aris, Hilal Ahmad, Tabassum Parveen, Jafariah Jaafar, Zaiton Abdul Majid, A. Vijaya Bhaskar Reddy, and Juhaizah Talib. "Synthesis of Visible Light Active Doped TiO_2 for the Degradation of Organic Pollutants—Methylene Blue and Glyphosate." *Journal of Analytical Science and Technology*, 7, no. 1 (2016). https://doi.org/10.1186/s40543-016-0109-2.

[68] Cornwell, Daniel J., and David K. Smith. "Expanding the Scope of Gels—Combining Polymers with Low-Molecular-Weight Gelators to Yield Modified Self-Assembling Smart Materials with High-Tech Applications." *Materials Horizons*, 2 (2015), 279–93. https://doi.org/10.1039/c4mh00245h.

[69] Feng, C., K. C. Khulbe, and T. Matsuura. "Recent Progress in the Preparation, Characterization, and Applications of Nanofibers and Nanofiber Membranes via Electrospinning/Interfacial Polymerization." *Journal of Applied Polymer Science*, 115, no. 2 (2010), 756–76. https://doi.org/10.1002/app.31059.

[70] Nemiwal, Meena, Tian C. Zhang, and Dinesh Kumar. "Graphene-Based Electrocatalysts: Hydrogen Evolution Reactions and Overall Water Splitting." *International Journal of Hydrogen Energy*, 46 (2021), 21401–18. https://doi.org/10.1016/j.ijhydene.2021.04.008.

[71] Yu, Li, Shuangchen Ruan, Xintong Xu, Rujia Zou, and Junqing Hu. "One-Dimensional Nanomaterial-Assembled Macroscopic Membranes for Water Treatment." *Nano Today*, 17 (2017), 79–95. https://doi.org/10.1016/j.nantod.2017.10.012.

[72] Shen, Yue xiao, Patrick O. Saboe, Ian T. Sines, Mustafa Erbakan, and Manish Kumar. "Biomimetic Membranes: A Review." *Journal of Membrane Science*, 454 (2014), 359–81. https://doi.org/10.1016/j.memsci.2013.12.019.

[73] Sahebi, Soleyman, Mohammad Sheikhi, and Bahman Ramavandi. "A New Biomimetic Aquaporin Thin Film Composite Membrane for Forward Osmosis: Characterization and Performance Assessment." *Desalination and Water Treatment*, 148 (2019), 42–50. https://doi.org/10.5004/dwt.2019.23748.

[74] Lee, X. J., L. Y. Lee, L. P.Y. Foo, K. W. Tan, and D. G. Hassell. "Evaluation of Carbon-Based Nanosorbents Synthesised by Ethylene Decomposition on Stainless Steel Substrates as Potential Sequestrating Materials for Nickel Ions in Aqueous Solution." *Journal of Environmental Sciences (China)*, 24, no. 9 (2012), 1559–68. https://doi.org/10.1016/S1001-0742(11)60987-X.

2 Advanced Carbon-Based Nanomaterials for Treatment of Water

Sapna Nehra,[1]* Sayed Zenab Hasan,[2]
Rekha Sharma,[3] and Dinesh Kumar[4]
[1]Department of Chemistry, Nirwan University, Jaipur, Rajasthan 303305, India
[2]Department of Chemistry, Dr. K. N. Modi University, Newai, Rajasthan 304021, India
[3]Department of Chemistry, Banasthali Vidyapith, Banasthali, Rajasthan 304022, India
[4]School of Chemical Sciences, Central University of Gujarat, Gandhinagar, India
*Corresponding author: Email: nehrasapna111@gmail.com

CONTENTS

2.1	Introduction	25
2.2	The Application of Single-Walled Carbon Nanotubes in Water Treatment	27
2.3	The Application of Multi-Walled Carbon Nanotubes in Water Treatment	29
2.4	The Application of Carbon Nanotube Composites	31
2.5	Graphene Oxide-Based Composites In Water Treatment	33
2.6	Conclusion	37
References		37

2.1 INTRODUCTION

Clean water is an extremely critical issue in developing countries. Around 40% population of the world cannot access clean water. In developing countries, faster growth of urbanization and industrialization has resulted in a freshwater crisis. A huge amount of water and other human resources is required to construct infrastructure. This boom in population has resulted in the overexploitation of resources. Universal freshwater problems are predominant, and therefore the need for freshwater is constant. So, researchers are looking towards the purification techniques of saline and waste water. Contamination of water leads to freshwater availability [1]. Carbon nanotubes (CNTs) are a new carbon family member discovered by Iijima in 1991.

CNTs have been the center of significant research due to their distinctive physicochemical properties [2–4]. Well-aligned CNTs could function as strong pores in membranes for water decontamination and desalination applications. The cavity of the CNT structure gives frictionless transfer of water molecules and makes them fit for the advancement of high fluxing separation technologies. A suitable pore size could account for the channel entries energy barriers, allowing water through the nanotube cavity and rejecting salt ions. Also, it is practicable to carefully reform CNT pores to sense and reject ions [5]. Predictably, the selective and fast fluid transfer seen in CNTs has quickly initiated much continuing scientific research and discussion to use this unique feature [6].

Carbon nanomaterials are turning into remarkable materials to treat water. We are gaining the advantages of enhanced adsorption efficiency, superior performance, and material use reduction in many current applications that the benefits of these advanced materials have allowed [7]. These nanomaterials have been universally used to prevent, remedy, and treat effluent. In addition, these biodegradable materials can conserve the feasibility and accessibility of water supply and long-term water quality while regularly detecting the chemical and biological pollutants from industrial, municipal, and man-made waste. These carbon-based nanomaterials with unique benefits could be considered to be good alternatives for waste water treatment [8].

To resolve the major issue of freshwater, scientists opted for several purification methods, such as using carbon material as CNTs. Many inorganic, dyes, biological, and organic pollutants can be targeted through the CNT [9-12]. Some properties like higher surface area, minimum cost, and that they are more reactive, make them commercially more suitable for purification. In past years, 736 tonnes of CNTs were utilized in energy and environmental applications [13-15]. Improved characteristics of carbon-based materials like carbon nanotubes, graphene oxide, and various types of CNT materials were produced with various polymers, ceramics, and the like. Several toxicants have been treated through nanomaterials synthesized from bare carbon material [16-19]. A series of contaminants found in waste water creates the need for a novel adsorbent that can further help target the multi-toxic ions. The significant pollutants in the aqueous system are inorganic ions, pesticides, heavy metal ions, pharmaceuticals, organic compounds, antibiotics, and so forth. The morphology of the advanced carbon composites, nanocomposites, and nanomaterials amplifies their properties, resulting in higher adsorption capacity, selectivity, and easy attainment of equilibrium and regeneration. The invention of efficient technologies permitting the selective capture of environmentally detrimental pollutants from waste water treatment plants and drinking water at affordable prices is attracting wide interest [20, 21].

This chapter emphasizes the applications of advanced sustainable nanomaterials to eliminate aqueous contaminants and pollutants by utilizing carbon-based nanomaterials, namely CNTs, including Single-walled carbon nanotubes

(SWCNTs), Multi-walled carbon nanotubes (MWCNTs), CNT composites, graphene, and its oxide.

2.2 THE APPLICATION OF SINGLE-WALLED CARBON NANOTUBES IN WATER TREATMENT

Carbon-based materials have been utilized as catalysts and support materials for the electrochemical sensing of gases and toxic chemicals. SWCNTs show unique optical and electrical properties, involving near-infrared (NIR) fluorescence, UV/visible/NIR absorption and large Raman scattering cross-sections [14]. A SWCNT can be explained as a single graphene sheet rolled into a tube. In addition, it is suggested that these tubes are either semiconducting or metallic, relying on their diameter and chirality (spiral conformation), to produce ideal reinforcing fillers in composite materials. These noteworthy features give rise to a good adsorption capacity [22].

Balarak et al. (2021) used SWCNTs to adsorb Acid Blue 92 (AB92). The physicochemical properties of the adsorbent were illustrated by using several analytical tools like FTIR, BET SEM, and TEM. The batch experiment was done to check the adsorption capacity of the adsorbent by varying the dosage amount, contact time, initial concentration of dye, and pH in different ranges such as 0.01 - 0.2 g/L, 10 - 180 mins, 10 - 200 mg/L, and 3 - 11 mins respectively. With the 0.12 g/L, 10 mg/L, 3, and 75-min dosage amounts, initial dye concentration, pH, and contact time achieved the maximum adsorption capacity. Pseudo-second-order kinetics and a Langmuir isotherm identified the best-fitting model in the adsorption experiment, and a monolayer type of maximum adsorption of 86.91 mg/g was achieved at 333 K. Further, 16.62 kJ/mol activation energy was reported. According to a thermodynamics study, ΔS_0 confirmed the affinity of adsorbent with AB92. Plenty of forces were responsible, including π - π acceptor–donor interactions, dipole-induced dipole bonds, London dispersion interactions, hydrogen bonds, and the hydrophobic effect. The obtained thermodynamic and adsorption results showed effective adsorption against AB92 removal from industrial effluents and waste water [23].

Fard et al. (2021) studied the ethylenediamine-surface-modified EDA-SWCNTs for the adsorption of Pb(II) and Cd(II) from waste water. Several techniques such as FTIR, FESEM, and EDX analyzed the EDA-SWCNTs. Modified CNT at 1 atm, enhanced the adsorption percentage of Pb(II), and Cd(II) was 52% and 72%, respectively. Various parameters were studied for the removal of Pb(II) and Cd(II), for instance, initial concentrations of Pb(II) and Cd(II), adsorbent dosage, temperature, contact time, and the effects of co-existing ions. Atomic absorption spectrophotometry gave an overall determination of the adsorption capacity. Central composite design modeling was used to optimize the adsorption results given optimum conditions. Subsequently, optimized results favored the initial Pb(II) concentration, Cd(II), contact time, pH, and dosage as essential parameters for the adsorption process. Similarly, the adsorption capacity for the Pb(II) and

Cd(II) was found to be 96.91% and 93.47%. Optimum adsorption was achieved at the 44 mg/L of the initial concentration of metallic ions, 50 mins contact time, 5 pH, and 200 mg of EDA-SWCNTs. Further, experimental resultant data was put through a validation test and found errors of 1.25% for Pb(II) and 1.95% Cd(II). The equilibrium isotherm was well fitted with the Langmuir adsorption isotherm model, indicating the homogenous monolayer adsorption. The obtained regression coefficient (R^2) value was observed at 0.999 and 0.997 for Pb(II) and Cd(II) [24].

Baratta et al. (2022) synthesized environmentally friendly SWCNT-BP augmented with MTV-MOF for the selective and efficient removal of Pb(II) from multicomponent water systems. Bare MTV-MOF was easily incorporated into a porous network of entangled SWCNTs, forming a self-standing adsorbing membrane filter (MTV-MOF/SWCNT-BP). Without immobilization, the single SWCNT-BP exhibited an excellent exclusion rate with a higher adsorption capacity of 180 mg/g. Other than adsorption, the selectivity also persists from 200 to 10,000 ppb for Pb(II) amongst the multi-ion solutions. Significantly, the immobilization features increase selectivity by 42% and adsorption capacity up to 310 mg/g compared to the bare SWCNT-BP. The carbon material-related membrane reduced the Pb(II) concentration from its critical level of 1000 to 10 ppb, inadequate limits set by EPA and WHO. Therefore, the environmentally friendly carbon composite has shown potential applications in Pb(II) decontamination from household drinking water and industrial treatment plants [25].

Arsalan et al. (2021) demonstrated the removal of ammonia pollutants. Ammonia acts as an environmental pollutant due to its odor, corrosion, and algae causing nature. They studied the removal of ammonia from waste water. Subsequently, Design-Expert software designed the adsorption experiment and checked the optimizing parameters such adsorbent dosage, temperature, ammonia concentration, and contact time. The obtained results confirmed that pH 9.5 attained the maximum adsorption. The adsorption rate enhanced the dosage amount and temperature. In addition, following the raising of the adsorption rate, the reduction follows the temperature level and the quantity of the adsorbate [26].

Fernandez et al. (2021) presented the removal of triclosan (TCS) by using SWCNTs. After the synthesis, the adsorbent was checked through several analytical tools such as SEM, TEM, and EDS analysis to determine the surface morphology. The working range pH was confirmed at the point of zero charges. During the experiment it was found that the pH of the solution affects the adsorption equilibrium. The kinetics and thermodynamic studies estimated that the whole adsorption occurred on the adsorbent surface. They implemented kinetic models and a diffusional model related to the pore volume diffusion. Maximum adsorption was attained at seven pH levels with 30.3 mg/g. Further, the kinetics model supported the pseudo-first-order model and the diffusional model [27].

Ashraf et al. (2021) initially took the number of CA/PVP ratios and used the two-stage phase inversion method to synthesize cellulose acetate poly(vinylpyrrolidone) (PVP) membranes. A mixture of a solution containing K(I), Na(I), Ni(II), and Zn(II) separated through the dialysis process from its pairs such as K(I)/Zn(II), K(I)/Ni(II), Na(I)/Ni(II) and Na(I)/Zn(II). The highest separation factor membrane was selected for further modification with a varying amount of 0.1 to 0.7 wt.% of acid f-SWCNTs. Results from the f-SWCNTs infused membranes were then checked for their pure water flux morphology, ion dialysis separation, hydrophilicity, contact angle, water uptake, and thermal stability. The interactions between f-SWCNTs and the membrane matrix were investigated through FTIR spectroscopy. The thermal stability of f-SWCNTs infused membranes was improved, which was confirmed by TGA thermograms. Surface morphological examination exposed asymmetric structures and uniform distributions of f-SWCNTs in the membranes. During the examination of water uptake and contact angle, the hydrophilicity of the modified membranes was raised. The pure water flux was diminished by raising the infusion of f-SWCNTs. On increasing the infusion, the membranes were found to have increased the ion separation factor, although the ion permeation flux deteriorated. The CPC3 and CPC4 membranes differ in terms of infusion as 0.5 wt.% and 0.7 wt.% f-SWCNTs showed a separation of the almost same ions. CPC3 was the best ion separating SWCNTs membrane [28].

Fu et al. (2022) treated high salinity waste water, which is quite an issue due to the quenching effect of many anions on radical processes. Nonradical processes may be more favorable. Here three-dimensional nitrogen-doped graphene nanosheet networks were impregnated with carbon nanotubes (N-GS-CNTs) followed by direct pyrolysis of $K_3Fe(CN)_6$. Further, the synthesized catalyst can successfully activate PMS for the degradation of tetracycline. This showed a broad pH range from 3 to 11, and even a high salinity water concentration of about 500 mM HCO_3^-, Cl^-, and many more ions. Both DFT calculations and experimental characterizations defined the degradation mechanism well. Good catalytic efficacy was accredited to an increased rate of electron transfer from donor tetracycline to acceptor PMS, in the presence of the catalyst that behaves as an electron shuttle mediator to boost the nonradical process. At the same time, the catalyst also improves the production of singlet oxygen ($^1O_2$), therefore further enhancing the degradation rate. So, this study gave the degradation path of organic pollutants even in a high salinity medium [29].

2.3 THE APPLICATION OF MULTI-WALLED CARBON NANOTUBES IN WATER TREATMENT

MWCNTs were the first observed CNTs, comprising up to many tens of graphitic shells and ~0.34 nm adjacent shell separation, ~1 nm diameters, and large length/diameter proportion. MWCNTs could be observed as elongated fullerene. Besides, MWCNTs have acquired certain recognition because of their large surface areas, high electrical conductivity, and chemical stability. Due to these characteristics, MWCNTs could be utilized as electrode support materials to improve the electrochemical performance and electron transfer rate [30].

Luca et al. (2021) employed MWCNTs in the waste water adsorption of organic textile dyes. They prepared an adsorbent column using the MWCNTs. Further, the column was optimized, and the discoloration of toxicants present in the aqueous system was checked. The concentration of Reactive Black 5 (RB5) 37 mg/L was maintained at the time of optimization. Further, the adsorption-desorption cycles were performed with a predetermined volume of 100 mL until a colorless eluted solution was obtained. The adsorption process was performed with various columns and found unique adsorption efficiency. Each column's efficiency depends on the available concentration of dye solution, empty bed contact time (EBCT), amount of MWCNTs, and carbon usage rate (CUR). The saturated solution was checked through UV-Visible and TOC analysis, but with the help of a thermogravimetric TG-DTA tool-analyzed synthesized material. Authors found that only 2.5 g of CNT containing columns out of other utilized columns showed better results towards Reactive Black 5. The obtained adsorption efficacy was demonstrated as 55 mg/g [31].

Chen et al. (2021) focused on a MWCNTs based adsorbent to remove TC and heavy metals from an aqueous system using a 0.15 g IL-MWCNTs composite tablet. Benzothiazole ionic liquid (*N*-butyl benzothiazole hexafluoroborate, $[C_4Bth][PF_6]$) showed fine selectivity, which was incorporated into MWCNTs before tableting under 5 - 20 MPa. The synthesis method was quite simple and exhibited a magnetic core of dimensions 0.6 mm × 8 mm, which attributed to dispersing the solution for accelerating mass transfer. Subsequently, the unique multi-walled composite tablet was analyzed via FESEM, TGA, Zeta potential, and spectral techniques. In addition, many parameters were studied to investigate its adsorption efficacy. The novel adsorbent was very suitable for the TC and the heavy metals, Cr(IV), and Cu(II) reached optimum rates of 99.76%, 94.10%, and 84.60%, respectively. PSO and Langmuir model with a higher regression value $R^2 \geq 0.99$ proved the well-fitted adsorption data. The used adsorbent was regenerated by performing the adsorption-desorption with 5 mol/L NaOH solution. They found that it could achieve recycled composite tables over 90% adsorbate [32].

Lee et al. (2022) prepared the novel magnetic MWCNTs to isolate saturated adsorbents from treated water. Here, MWCNTs mixed with Fe_3O_4 were subsequently fixed with PDDA and finally formed PDDA-MMWCNTs by a solvent-free direct doping process. Nitric acid was in the pre-treatment process before mixing bare MWCNTs. Remarkable techniques like XRD, EDX, FTIR, and zeta potential confirmed the doping of iron NPs over the MWCNTs. It estimated the solidity of the synthesized MMWCNTs in the presence of given different pH levels under different conditions. Correspondingly, the magnetite coated over the n-MMWCNTs and PDDA-MMWCNTs separated at pH 9 and pH 3. Subsequently, these types of MMWCNTs were employed for the exclusion of MB. PDDA-MMWCNTs demonstrated more adsorption capacity than the MMWCNTs. This was 11% more due to the remarkable higher specific surface area of n-MMWCNTs.

Further, in pH range 3 - 11, the adsorption rate rise by around 87%, indicating that the alkaline condition is more favorable for the adsorption process. The obtained adsorption results were better fit to the Langmuir isotherm, which signifies a monolayer type of adsorption of MB with higher efficiency of 20.37 mg/g. Further, the pseudo-second-order kinetic model followed the MB adsorption [33].

2.4 THE APPLICATION OF CARBON NANOTUBE COMPOSITES

The extraordinary physical and mechanical properties of carbon nanotubes, combined with their low density, makes this new form of carbon a viable choice for composite reinforcement. Significant basic research is required before these exceptional properties noticed at the nanoscale are conceived in a macroscopic composite. Complete consideration of the thermo-mechanical property of nanotube-based composites needs an understanding of the fracture and elastic properties of carbon nanotubes and their relationships at the nanotube/matrix interface. In addition, the atomic structure of the nanotubes is expressed in terms of the tube helicity or chirality, which is explained by the chiral angle and the chiral vector. Advanced research in nanotube-based composites has shown the capability of carbon nanotubes for reinforcement [34].

Yadav et al. (2022) opted for a novel method to develop the CNT impregnated activated carbon procured from the eucalyptus. The developed carbon-related adsorbent was used to absorb two different dyes, EY and MB, from the waste water. Various parameters were checked to investigate the efficacy of the adsorbent, such as the effect of initial concentration, adsorbent dose, pH, and contact time. Obtained adsorption data were fitted with many isotherms like the Langmuir isotherm, the Freundlich isotherm, and the D-R isotherm. Amid all isotherm models, the D-R model was the most favorable and best fit. They reported that the higher adsorption efficiency toward the EY and MB was 49.15 and 49.61 mg/g. Besides thermodynamics and kinetics, studies were done to understand the toxic dye adsorption mechanism better. During kinetics examination, it found the PSO model validated the adsorption results. Therefore, eucalyptus integrated CNTs were shown to have good applicability for removing dyes from waste water [35].

Arabkhani et al. (2021) synthesized the carbon nanocomposite adsorbent by incorporating magnetic tungsten disulfide designated as $WS_2/Fe_3O_4/CNTs$-NC. The adsorbent was further used to exclude the AM and BB FCF dyes. Results were optimized by using the central CCD method, giving the inputs in the form X1, X2, X3, X4, and X5, which represented the pH, AM concentration, BB FCF concentration, adsorbent mass, and sonication time, which were, 2.0 - 10, 10 - 50 10 - 50 mg/L, 4 - 20 mg, and 2 - 12 min correspondingly. At optimum conditions, the exclusion rate of the dye was obtained at 99.30% and 98.50% for AM and BB FCF. Thermodynamics and kinetic studies were estimated in Langmuir isotherm and PSO models. The optimum adsorption capacity was 174.8 and 166.7 mg/g for AM and BB FCF. A thermodynamics study revealed that the adsorption of dyes occurred endothermically and spontaneously. Using real water, the analysis found quite a good adsorption percentage of 94.52 - 99.65% of AM and BB FCF in the binary solution. Even after five consecutive adsorption-desorption cycles, it still maintained an adsorption rate of over 90% for each dye. So, this study attributed a new insight into the potential application of magnetic carbon nanocomposite for waste water treatment with dyes [36].

Jin et al. (2021) reported the synthesis of facile incorporated magnetic CNT/chitosan composite to treat manifold contaminants through a two-step method, which displayed good adsorption efficiency of organic toxins and heavy metals. Because of its magnetic nature, the adsorbed composite can be isolated by applying an external magnetic field. A significant analysis was used to determine the interaction within the functional groups, structural formations, surface morphologies, stabilities, and magnetic natures such as FTIR, XRD, TEM, FESEM, TGA, and VSM, respectively. The obtained composite displayed excellent adsorption of organic Rhodamine B compared to methyl orange because of the strong π-π electron-donor-acceptor interface among the benzene ring of RhB and CNT. The adsorption rate for heavy metals trailed the sequence as $Fe(III) > Cu(II) > Pb(II) > Zn(II)$. The PSO model was suitable and kinetically preferable for heavy metals and organic pollutants. The kinetics result this preference because of the synergy of the π–π stacking and hydrophobic CNT and chelation interaction of chitosan. In addition, spent adsorbent could be regenerated and exhibited a good adsorption capacity, symbolizing the best adsorption to deal with the treatment of multiple in aqueous systems [37].

Marszałek et al. (2022) chose a very adaptable adsorbent that simultaneously eliminated organic and inorganic multi-pollutants found in rainwater. This work evaluated natural, temperature-dried bentonites from 0 to 700°C attributed to the Ben100 + SWCNT and Ben100 + SWCNT-OH. In the formation of Ben100 the ratio of 60:40 was taken, which means 60 wt.% of bentonite and 40 wt.% of SWCNT. The rainwater configuration was reformed by adding lead nitrate, zinc nitrate, and additional organic micropollutants collected with ANT and BZT. The elimination of BZT and ANT ranged from 4.5% to 93% through Ben100 + SWCNT and from 50% to 95.5%, through Ben100 + SWCNT-OH.

Ben100 + SWCNT showed 93% exclusion towards BZT, while Ben100 + SWCNT-OH displayed 95.5% for ANT. Between the Freundlich and the Langmuir isotherm models, the Freundlich suited adsorption on Ben100 and Ben100 + SWCNT well. Moreover, the Freundlich model favored adsorption on Ben100 + SWCNT-OH for BZT and ANT. A thermal range from 0 to 700°C bentonite (Ben100 and Ben300) demonstrated maximum removal of both $Zn(II)$ and $Pb(II)$ and a minimum removal by Ben500 and Ben700. The 90% elimination percentage was achieved through Ben100 + SWCNT and Ben100 + SWCNT-OH composites for $Zn(II)$ and $Pb(II)$. Both composites attained equilibrium in a very short time frame of approximately 30 min. Therefore, all obtained experimental results indicated that Ben-CNT possessed potential application in the elimination of micropollutants together with heavy metals from aqueous systems [38].

Yang et al. (2021) employed the CNT background adsorbents to remove the number of hazardous dyes found in aqueous systems. The carbon-related adsorbent- fronting blockades increased rejuvenation but lowered adsorption efficiency. Therefore, a huge adsorptive filter was prepared using photo-regenerable carbon nanotubes that possessed good self-cleaning assets. Further, the facile vacuum filtration method, trialed by FeOOH in-situ anchoring and silver-amino reaction. The synthesized carbon filtration membrane removed the toxic dyes and many pollutants in acid orange 7, rhodamine B, methylene blue, eosin Y, p-nitrophenol bisphenol A, and amoxicillin. In literature, reports have found that CNT membranes were more supportive of the removal of dyes than the other carbon-based adsorbents. The carbon-related composite filters exhibited excellent adsorption concerning cationic dyes.

The developed CNT film displayed the maximum adsorption capacity for methylene blue and rhodamine B, 247 and 181 mg/g. Even after significant organic fouling, it maintained very good self-cleansing qualities. Further, diverse functional carbon nanotube membranes demonstrated unique properties of catalytic degradation, membrane separation, and adsorption into a single system. So, it showed immense potential in regenerable, auto cleaning, and higher absorptivity for the adsorption of dyes from contaminated water [39].

Tang et al. (2021) focused on the new variety of pollution that spreads in microplastics (MPs). The basins made from MP had a severe negative impact on the sustainable environment and human health. Therefore, the efficient and environmentally friendly method was best. Firstly, in this study, they synthesized magnetic carbon nanotubes designated as M–CNTs. Subsequently, the MPs/M–CNTs effectively adsorbed the PA, PET, and PE. With the help of a magnetic field, the composite used could be extracted from the MPs/M–CNTs. The optimum adsorption condition was demonstrated with the initial concentration of the magnetic solution as 5 g/L and a contact time of 300 min. The maximum adsorption capacity was reported as 1650, 1400, and 1100 mg/g.

The adsorption efficacy was not affected by other interfering substances like COD, PO_4^{3-} and NH_3-N. All MPs were eliminated from the waste water discharged from a kitchen waste treatment plant. Further, the spent M–CNTs might be rejuvenated through thermal treatment at 600°C. The recycled M–CNTs similarly behaved like the original ones regarding adsorption capacity. Even after performing four repetitive cycles, it still showed approximately 80% removal of total MPs available in the optimized medium. To control microplastic pollution, the M–CNTs seemed to be a very effective and unique approach to treat waste water [40].

2.5 GRAPHENE OXIDE-BASED COMPOSITES IN WATER TREATMENT

Geim and Novoselov discovered graphene in 2010, winning the Nobel prize. The potential for graphene and graphene oxide (GO) for membrane construction has been broadly studied, mostly for GO nanosheets that could be easily constructed through the chemical exfoliation of graphite powders [41].

Han et al. (2021) showed that graphene and its derivatives have good application in waste water treatment and in the desalination of seawater because of their distinctive pore sizes and ionic molecular strain separation capabilities. GO, graphene, and rGO can be developed into composites and nanoporous materials with modulated properties that could be improved for water filtration. Techniques for penetrating graphene comprise electron beam nanometer engraving and ion bombardment. Graphene-based composites facilitate the enhancement of the potential of graphene in waste water treatment and in the desalination of seawater by utilizing its latest properties and features. This chapter focuses on enhancing the production of graphene-based membranes in separation, seawater desalination and decontamination applications, by examining how several alteration and construction processes affect major performance properties, such as water permeance, the declining of solutes, selectivity, antifouling characteristics, as well as membrane mechanical strength [42].

Hu et al. (2021) prepared a nickel nanoparticle (NiNP) enhanced graphene oxide-carbon nanotube nanocomposite by a new molecular-level-mixing process, then a freeze-drying and successive reduction procedure. The resulting products displayed a well-dispersed 3D structure. They successfully eliminated Rhodamine B (RhB) from the water system by the synergistic effect of photo-degradation and physical adsorption. Nanocomposite physical adsorption of RhB followed the pseudo-second order and this can be attributed to its large surface area. 30 wt.% of Ni nanocomposite indicated the excessive removal capability of approximately 41.5 mg/g in a solution of 5 ppm RhB. Besides the ferromagnetism of coated NiNPs, the nanocomposite can be collected and recycled after dye removal. Hence, the nanocomposite has shown good photocatalytic performance for removing organic dyes from waste water [43].

Ismail et al. (2021) prepared a nanohybrid of quaternized polydopamine reduced graphene oxide (QSiPD-rGO) by a one-step mussel-inspired modification of GO using dopamine and a quaternized silica precursor. The QSiPD-rGO showed a remarkable loading rate of 8 wt.% to produce positively charged, fouling resistant, antibacterial, and high flux hybrid polyethersulfone (PES) ultrafiltration (UF) membranes by non-solvent induced phase dissociation. A suite of tools was used to characterize the membranes and nanohybrid membranes, including Raman, SEM-EDX, ATR-FTIR, AFM, TGA, contact angle measurements, and zeta potential. The charge, hydrophilicity, and surface morphologies of membranes were regulated by combining the nanohybrids. Hence, pure water flux was remarkably enhanced, attaining ~ 270 L/m^2/h (at 1 bar) for PMQ4 with six wt.% QSiPD-rGO loadings.

This was 94% more than the equaling flux of the pristine membrane. The hybrid membranes exhibited superior fouling resistances during the UF of 500 ppm bovine serum albumin (BSA) solution. Thus, the hybrid membranes' BSA rejection rate was much greater and rigid for their high flux, surpassing 98% for PMQ4. In addition, the nanohybrid produced a significant antibacterial activity against *E. coli* in the membranes, with a strong interaction between this activity and QSiPD-rGO loading [44]. Miscellaneous advanced carbon nanomaterials are shown in Table 2.1.

TABLE 2.1
Distinct Types of Advance Carbon Nanomaterials

S. No.	Carbon-based Nanomaterials	Target Pollutant	Degradation/ Adsorption Efficiency	References
1.	GO and graphitic carbon nitride (g-C_3N_4)	Methylene blue, and tetracycline	220 and 120 mg/g	[45]
2.	Titanium doped activated carbon-cellulose nanocomposite	Crystal violet, Methyl violet	96%	[46]
3.	MWCNTs supported by zinc-oxide nanoparticle	Methylene blue	96.7%	[47]
4.	Chitosan modified nitrogen-doped porous carbon composite	Phenol, BPA and 2,4-DCP	254.45, 675.68 and 892.86 mg/g	[48]
5.	Cobalt-seamed metal–organic nanocapsule (Co-MONC),	Chlorpromazine (CPZ)	1039.53 mg/g	[49]
6.	Ag_2O/Bi_2WO_6 heterostructure incorporated reduced graphene oxide (ABW-rGO) composite	Tetracycline and Crystal violet (cationic dye) and Congo red (anionic dye)	95.3% and 98.5%	[50]
7.	Metal-organic framework/graphene oxide composite	Acid blue 92	371.8	[51]
8.	rGO and g-C_3N_4 nanosheets	RB5 azo dye	4.78 kg/m²/h	[52]
9.	Zeolitic Imidazolate framework/graphene oxide composites (ZIF-67@GO).	Malachite green	99.18%	[53]
10.	GO decorated with manganese ferrite	Malachite green and methylene blue dye	156, and 105 mg/g	[54]
11.	Novel PVC/PPD/GO and PVC/GO composite	Pb(II), Cu(II), Ni(II), Co(II), Zn(II) and Cd(II)	17.61, 22.17, 25.41, 33.57 and 44.80 mg/g	[55]

(*continued*)

TABLE 2.1 (Continued)
Distinct Types of Advance Carbon Nanomaterials

S. No.	Carbon-based Nanomaterials	Target Pollutant	Degradation/ Adsorption Efficiency	References
12.	MCS/GO-PEI	As(III), Hg(II), and congo red	220.26, 124.84, 162.07, 93.81 mg/g	[56]
13.	Gd_2O_3-GO@SA-CMC granules	Cr(III), Pb(II), and As(V)	29.16, 158.73, and 36.77 mg/g	[57]
14.	Gd_2O_3-doped graphene oxide	Cr(III) and Pb(II)	17.97 and 83.04 mg/g	[58]
15.	GO-thiosemicarbazide (mGO-TSC)	Cu(II) and Pb(II)	85% and 81.7%	[59]
16.	Zwitterionic GO-BPED-PS adsorbent	Ni(II), Co(II)	4.174 ± 0.098 mmol/g 3.902 ± 0.092 mmol/g	[60]
17.	Graphene oxide-Nickel oxide (GO-NiO)	Pb(II) and Cd(II)	208.8 and 324.7 mg/g	[61]
18.	Amine functionalized GO/MnO_2 nanohybrid	Cu(II), Zn(II), and Ni(II) ions	70.1, 49.1, and 55.8%	[62]
19.	Sand enriched GO	Cr(VI), As(III), Cd(II), and Pb(II)	530.85, 170.1, 49.78, and 14.41 mg/g	[63]
20.	Functionalized magnetic graphene oxide	Pb(II)	40 mg	[64]
21.	Natural bionanopolymer based on Schiff base chitosan/graphene oxide	lead and copper	466 and 698.8 mg/g	[65]
22.	EDTA-functionalized graphene oxide-chitosan nanocomposite	Hg(II), Cu(II), MB, and CV,	324±3.30, 130±2.80, 141±6.60, and 121±3.50 mg/g	[66]
23.	Magnetic fungal hyphal/ graphene oxide nanofibers (MFHGs)	Co(II) and Ni(II)	104.34, 97.44 mg/g	[67]
24.	Mg-Al LDH/C-dots nanocomposite.	Pb(II)	41.32 mg/g	[68]
25.	Fe_3O_4@C-GO	Ag(I), Pb(II), Cr(VI) and Al(III),	162.9, 125.8 158.2 and 173.9 mg/g	[69]

2.6 CONCLUSION

The treatment of water from industry is very important for protecting human health and the environment, as well as great concern in terms of public health. Carbon-based nanomaterials and the application of their nanocomposites in the treatment of water systems is essential to address the limitations of traditional treatment processes. Various investigations involving toxicity inspection, strategy assessment, the dispersion of NPs, and life cycle evaluation in water systems need to be conducted to examine the hazards of nanomaterials to human health. The conclusions of these would prove instructive and would guide good recognition of the conduct of NPs such as CNTs in water systems. Apart from CNT-based nanoadsorbents, other allotropes of nanocarbon, such as graphene, GO, or rGO, have exhibited the ability to improve the photocatalytic activities of traditional semiconductors. In addition, the high adsorption abilities of carbon nanomaterials could hold chemical pollutants that might appear, and which frequently have a negligible and weak attraction for traditional sorbents. So, these twofold performances may open the path for constructing hybrid water treatment techniques. Although, for the efficient commercialization of these techniques, considerable advancement would be required for extensive production and cost of these materials. Hence, research on the latest techniques to mass-produce carbon-based nanomaterials economically would be essential. Future studies are needed to concentrate on authenticating the efficacy of these nanocomposites and nanomaterials with natural water systems to address the effects of a range of environmental factors. Ultimately, special attention should be given to the construction of carbon-based nanomaterials with the least ecological and human impact for sustainable water treatment applications.

REFERENCES

[1] Sui, M. Zhang, L. Sheng, L. Huang, S. and She, L. "Synthesis of ZnO coated multi-walled carbon nanotubes and their antibacterial activities." *Science of the Total Environment* 452 (2013), 148–154. doi.org/10.1016/j.scitotenv.2013.02.056.

[2] Wang, S. Ng, C. Wang, W. Li, Q. and Zhengping, H. "Synergistic and competitive adsorption of organic dyes on multi-walled carbon nanotubes." *Chemical Engineering Journal* 197 (2012), 34–40. 0doi.org/10.1016/j.cej.2012.05.008.

[3] Keller, A. A. McFerran, S. Lazareva, A. and Suh, S. "Global life cycle releases of engineered nanomaterials." *Journal of Nanoparticle Research* 15, no. 6 (2013), 1–17. doi.org/10.1007/s11051-013-1692-4.

[4] Ren, X. Chen, C. Nagatsu, M. and Wang, X. "Carbon nanotubes as adsorbents in environmental pollution management: a review." *Chemical Engineering Journal* 170, no. 2–3 (2011), 395–410. doi.org/10.1016/j.cej.2010.08.045.

[5] Das, R. Ali, Md. E. A. Hamid, Sharifah B. A. H. Ramakrishna, S. and Chowdhury, Z. Z. "Carbon nanotube membranes for water purification: a bright future in water desalination." *Desalination* 336 (2014), 97–109. doi.org/10.1016/j.desal.2013.12.026.

[6] Goh, P. S. Ismail, A. F. and Ng, B. C. "Carbon nanotubes for desalination: Performance evaluation and current hurdles." *Desalination* 308 (2013), 2–14. doi.org/10.1016/j.desal.2012.07.040.

[7] Abbassian, K. Kargari, A. and Kaghazchi, T. "Phenol removal from aqueous solutions by, a novel industrial solvent." *Chemical Engineering Communications* 202 (2015), 408–413. doi.org/10.1080/00986445.2013.848804.

[8] Nasrollahzadeh, M. Sajjadi, M. Iravani, S. and Varma, R. S. "Carbon-based sustainable nanomaterials for water treatment: state-of-art and future perspectives." *Chemosphere* 263 (2021), 128005. doi.org/10.1016/j.chemosphere.2020.128005.

[9] Al-Muhtaseb, A. H. Ibrahim, K. A. Albadarin, A. B. Ali-Khashman, O. Walker, G. M. and Ahmad, M. N. M. "Remediation of phenol-contaminated water by adsorption using poly(methyl methacrylate) (PMMA)." *Chemical Engineering Journal* 168 (2011), 691–699. doi.org/10.1016/j.cej.2011.01.057.

[10] Ali, H. Khan, E. and Ilahi, I. "Environmental chemistry and ecotoxicology of hazardous heavy metals: environmental persistence, toxicity, and bioaccumulation." *Journal of Chemistry* 2019 (2019), Vol. 2019, pages, 1-14. doi.org/10.1155/2019/6730305.

[11] Banerjee, A. Gokhale, R. Bhatnagar, S. Jog, J. Bhardwaj, M. Lefez, B. Hannoyer, B. and Ogale, S. "MOF derived porous carbon-Fe3O4 nanocomposite as a high performance, recyclable environmental superadsorbent." *Journal of Materials Chemistry* 22, (2012), 19694–19699. doi.org/10.1039/c2jm33798c.

[12] Calcagnile, P. Fragouli, D. Bayer, I. S. Anyfantis, G. C. Martiradonna, L. Cozzoli, P. D. Cingolani, R. and Athanassiou, A. "Magnetically driven floating foams for the removal of oil contaminants from water." *ACS Nano* 6, no. 6 (2012), 5413–5419. doi.org/10.1021/nn3012948.

[13] Choi, J. H. Nguyen, F. T. Barone, P. W. Heller, D. A. Moll, A. E. Patel, D. Boppart, S. A. and Strano, M. S. "Multimodal biomedical imaging with asymmetric single-walled carbon nanotube/iron oxide nanoparticle complexes." *Nano Letters* 7 (2007), 861–867. doi.org/10.1021/nl062306v.

[14] Chu, Y. and Pan, Q. "Three-dimensionally macroporous Fe/C nanocomposites as highly selective oil-absorption materials." *ACS Applied. Material Interfaces* 4 (2012), 2420–2425. doi.org/10.1021/am3000825.

[15] Dabrowski, A. Podkościelny, P. Hubicki, Z, and Barczak, M. "Adsorption of phenolic compounds by activated carbon—a critical review." *Chemosphere* 58 (2005), 1049–1070. doi.org/10.1016/j.chemosphere.2004.09.067.

[16] Damjanovic, L. Rakic, V. Rac, V. Stosic, D. and Auroux, A. "The investigation of phenol removal from aqueous solutions by zeolites as solid adsorbents." *Journal of Hazardous Material* 184 (2010), 477–484. doi.org/10.1016/j.jhazmat.2010.08.059.

[17] Demir, A. Topkaya, R. and Baykal, A. "Green synthesis of superparamagnetic Fe3O4 nanoparticles with maltose: its magnetic investigation." *Polyhedron* 65 (2013), 282–287. doi.org/10.1016/j.poly.2013.08.041.

[18] Duan, J. Chen, S. Jaroniec, M. and Qiao, S. Z. "Heteroatom-doped graphene-based materials for energy-relevant electrocatalytic processes." *ACS Catal.* 5, no. 9 (2015), 5207–5234. doi.org/10.1021/acscatal.5b00991.

[19] Naik, S. S. Lee, S. J. Yeon, S. Yu, Y. and Choi, M. Y. "Pulsed laser-assisted synthesis of metal and nonmetal-codoped ZnO for efficient photocatalytic degradation of Rhodamine B under solar light irradiation." *Chemosphere* 274 (2021b), 129782. doi.org/10.1016/j.chemosphere.2021.129782.

[20] Liu, J. Morales-Narvaez, E. Vicent, T. Merkoci, A. and Zhong, G-H. "Microorganism-decorated nanocellulose for efficient diuron removal." *Chemical Engineering Journal* 354 (2018), 1083–1091. doi.org/10.1016/j.cej.2018.08.035.

[21] Jain, N. and Kanu, N. J. "The potential application of carbon nanotubes in water Treatment: a state-of-the-art-review." *Materials Today: Proceedings* 43 (2021), 2998–3005. doi.org/10.1016/j.matpr.2021.01.331.
[22] Kymakis, E. Alexandou, I. and Amaratunga, G. A. J. "Single-walled carbon nanotube–polymer composites: electrical, optical and structural investigation." *Synthetic Metals* 127, no. 1–3 (2002), 59–62. doi.org/10.1016/S0379-6779(01)00592-6.
[23] Balarak, D. Zafariyan, M. Igwegbe, C. A. Onyechi, K. K. and Ighalo, J. O. "Adsorption of acid blue 92 dye from aqueous solutions by single-walled carbon nanotubes: isothermal, kinetic, and thermodynamic studies." *Environmental Processes* 8, no. 2, (2021), 869–888. doi.org/10.1007/s40710-021-00505-3.
[24] Fard, E. M. Parvareh, A. and Moravaji, M. K. "Optimization of removal of lead and cadmium from industrial waste water by ethylenediamine-modified single-walled carbon nanotubes." *International Journal of Environmental Science and Technology* 19 (2022), 2747–2760. doi.org/10.1007/s13762-021-03390-3.
[25] Baratta, M. Mastropietro, T. F. Bruno, R. Tursi, A. Negro, C. Ferrando-Soria, J. Mashin, A. I. Nezhdanov, A. Nicoletta, F. P. de Filpo, G. Pardo, E. and Armentano, D. "Multivariate metal–organic framework/single-walled carbon nanotube buckypaper for selective lead decontamination." *ACS Applied Nano Materials* (2022). Vol 5, 5223–5233. doi.org/10.1021/acsanm.2c00280.
[26] Arsalan, J. Soheila, R. Ghasem, H. Roohullah, J. and Hossein, M. "Optimization and modeling of ammonia removal from aqueous solutions by using adsorption on single-walled carbon nanotubes." *Journal of Engineering & Technological Sciences* 53, no. 3 (2021). doi.org/10.5614/j.eng.technol.sci.2021.53.3.9
[27] Gonzalez-Fernandez, L. A. Medellín-Castillo, N. A. Ocampo-Pérez, R. Hernández-Mendoza, H. Berber-Mendoza, M. S. and Aldama-Aguileraa, C. "Equilibrium and kinetic modelling of triclosan adsorption on Single-Walled Carbon Nanotubes." *Journal of Environmental Chemical Engineering* 9, no. 6 (2021), p106382. doi.org/10.1016/j.jece.2021.106382.
[28] Ashraf, M. A. Islam, A. Dilshad, M. R. Butt, M. A. Jamshaid, F. Ahmad, A. and Khan, R. U. "Synthesis and characterization of functionalized single walled carbon nanotubes infused cellulose acetate/poly (vinylpyrrolidone) dialysis membranes for ion separation application." *Journal of Environmental Chemical Engineering* 9, no. 4 (2021), 105506. doi.org/10.1016/j.jece.2021.105506.
[29] Fu, C. Sun, G. Wang, C. Wei, B. Ran, G. and Song, Q. "Fabrication of nitrogen-doped graphene nanosheets anchored with carbon nanotubes for the degradation of tetracycline in saline water." *Environ. Res.* 206,(2022), Vol 5, 112242. doi:10.1016/j.envres.2021.112242.
[30] Naik, S. S. Lee, S. J. Theerthagiri, J. Yu, Y. and Choi, M. Y. "Rapid and highly selective electrochemical sensor based on ZnS/Au-decorated f-multi-walled carbon nanotube nanocomposites produced via pulsed laser technique for detection of toxic nitro compounds." *Journal of Hazardous Material* 418 (2021), 126269. doi.org/10.1016/j.jhazmat.2021.126269.
[31] de Luca, P. Chiodo, A. Macario, A. Carlo, S. and Nagy, B. J. "Semi-continuous adsorption processes with multi-walled carbon nanotubes for the treatment of water contaminated by an organic textile dye." *Applied Science* 11, no. 4 (2021), 2076–3417. doi.org/10.3390/app11041687.
[32] Chen, C. Feng, X. and Yao, S. "Ionic liquid-multi walled carbon nanotubes composite tablet for continuous adsorption of tetracyclines and heavy

metals." *Journal of Cleaner Production* 286 (2021), 124937. doi.org/10.1016/j.jclepro.2020.124937.

[33] Lee, C. S. Shuit, S. H. Lim, C. C. Ng, Q. H. Hoo, P. Y. Lim, S. and Teoh, Y. P. "Synthesis of magnetic multi-walled carbon nanotubes via facile and solvent-free direct doping method for water remediation." *Journal of Water Process Engineering* 45, (2022), 102487. doi.org/10.1016/j.jwpe.2021.102487.

[34] Thostenson, E. Ren, Z. and Chou, T-W. "Advances in the science and technology of carbon nanotubes and their composites: a review." *Composites Science and Technology* 61, no. 13 (2001), 1899–1912. doi.org/10.1016/S0266-3538(01)00094-X.

[35] Yadav, S. K. Dhakate, S. R. and Singh, B. P. "Carbon nanotube incorporated eucalyptus derived activated carbon-based novel adsorbent for efficient removal of methylene blue and eosin yellow dyes." *Bioresource Technology* 344 (2022), 126231. doi.org/10.1016/j.biortech.2021.126231.

[36] Arabkhani, P. Hamedrez, A. Asfaram, A. Sadeghfar, F. and Sadegh, F. "Synthesis of magnetic tungsten disulfide/carbon nanotubes nanocomposite (WS_2/Fe_3O_4/CNTs-NC) for highly efficient ultrasound-assisted rapid removal of amaranth and brilliant blue FCF hazardous dyes." *Journal of Hazardous Materials* 420 (2021), 126644. doi.org/10.1016/j.jhazmat.2021.126644.

[37] Jin, J. Sun, J. Lv, K. Huang, X. Wang, J. Jingping, L. Bai, Y. Guo, X. Zhao, J. Liu, J. and Hou, Q. "Magnetic-responsive CNT/chitosan composite as stabilizer and adsorbent for organic contaminants and heavy metal removal." *Journal of Molecular Liquids* 334 (2021), 116087. doi.org/10.1016/j.molliq.2021.116087.

[38] Marszałek, A. Kamińska, G. and Salam, N. F. A. "Simultaneous adsorption of organic and inorganic micropollutants from rainwater by bentonite and bentonite-carbon nanotubes composites." *Journal of Water Process Engineering* 46 (2022), 102550. doi.org/10.1016/j.jwpe.2021.102550.

[39] Yang, Y. Xiong, Z. Wang, Z. Liu, Y. He, Z. Cao, A. Zhu, L. and Zhao, S. "Super-adsorptive and photo-regenerable carbon nanotube-based membrane for highly efficient water purification." *Journal of Membrane Science* 621 (2021), 119000. doi.org/10.1016/j.memsci.2020.119000.

[40] Tang, Y. Zhang, S. Su, Y. Wu, D. Zhao, Y. and Xie, B. "Removal of microplastics from aqueous solutions by magnetic carbon nanotubes." *Chemical Engineering Journal* 406 (2021), 126804. doi.org/10.1016/j.cej.2020.126804.

[41] Wu, W. Shi, Y. Liu, G. Fan, X. and Yu, Y. "Recent development of graphene oxide based forward osmosis membrane for water treatment: a critical review." *Desalination* 491 (2020), 114452. doi.org/10.1016/j.desal.2020.114452.

[42] Han, Z-Y. Huang, L-J. Qu, H-J. Wang, Y-X. Zhang, Z-J. Rong, Q. L. Sang, Z-Q. Wang, Y. Kipper, M. J. and Tang, J-G. "A review of performance improvement strategies for graphene oxide-based and graphene-based membranes in water treatment." *Journal of Materials Science* 56, no. 16 (2021), 9545–9574. doi.org/10.1007/s10853-021-05873-7.

[43] Hu, C. Le, A. T. Pung, S. Y. Stevens, L. Neate, N. Hou, X. Grant, D. and Xu, F. "Efficient dye-removal via Ni-decorated graphene oxide-carbon nanotube nanocomposites." *Materials Chemistry and Physics* 260 (2021), 124117. doi.org/10.1016/j.matchemphys.2020.124117.

[44] Ismail, R. A. Kumar, M. Thomas, N. An, A. K. and Arafat, H. A. "Multifunctional hybrid UF membrane from poly (ether sulfone) and quaternized polydopamine

anchored reduced graphene oxide nanohybrid for water treatment." *Journal of Membrane Science* 639 (2021), 119779. doi.org/10.1016/j.memsci.2021.119779.

[45] Ku Sahoo, S. Padhiari, S. Biswal, S. K. Panda, B. B. and Hotaa, G. "Fe3O4 nanoparticles functionalized GO/g-C3N4 nanocomposite: an efficient magnetic nanoadsorbent for adsorptive removal of organic pollutants." *Materials Chemistry and Physics* 244 (2021), 122710. doi.org/10.1016/j.matchemphys.2020.122710.

[46] Maqbool, Q. Barucca, G. Sabbatini, S. Parlapiano, M. Ruello, M. L. and Tittarelli, F. "Transformation of industrial and organic waste into titanium doped activated carbon—cellulose nanocomposite for rapid removal of organic pollutants." *Journal of Hazardous Materials* 423, (2022). 126958. doi.org/10.1016/j.jhazmat.2021.126958.

[47] Bhuvaneswari, K. Palanisamy, G. Sivashanmugan, K. Pazhanivel, T. and Maiyalagan, T. "ZnO nanoparticles decorated multiwall carbon nanotube assisted ZnMgAl layered triple hydroxide hybrid photocatalyst for visible light-driven organic pollutants removal." *Journal of Environmental Chemical Engineering* 9, no. 1 (2021), 104909. doi.org/10.1016/j.jece.2020.104909.

[48] Liu, Y. Li, L. Duan, Z. You, Q. Liao, G. and Wang, O. "Chitosan modified nitrogen-doped porous carbon composite as a highly-efficient adsorbent for phenolic pollutants removal." *Colloids and Surfaces A: Physicochemical and Engineering Aspects* 610 (2021), 125728. doi.org/10.1016/j.colsurfa.2020.125728.

[49] Pan, Y. Rao, C. Tan, X. Ling, Y. Singh, A. Kumar, A. Li, B. and Liu, J. "Cobalt-seamed C-methylpyrogallol[4]arene nanocapsules-derived magnetic carbon cubes as advanced adsorbent toward drug contaminant removal." *Chemical Engineering Journal* 433, no. 3 (2022), 133857. doi.org/10.1016/j.cej.2021.133857.

[50] Wei, J. Chen, Z. and Tong Z. "Engineering Z-scheme silver oxide/bismuth tungstate heterostructure incorporated reduced graphene oxide with superior visible-light photocatalytic activity". *Journal of Colloid and Interface Science* 596 (2021), 22–33. doi.org/10.1016/j.jcis.2021.03.117.

[51] Hoseinzadeh, H. Hayati, B. Ghaheh, F. S. Kumars, S. S. and Mahmoodi, N. M. "Development of room temperature synthesized and functionalized metal-organic framework/graphene oxide composite and pollutant adsorption ability." *Materials Research Bulletin* 142 (2021), 111408. doi.org/10.1016/j.materresbull.2021.111408.

[52] Karimi-Nazarabad, M. Goharshadi, E. K. Mehrkhah, R. and Davardoostmanesh, M. "Highly efficient clean water production: reduced graphene oxide/graphitic carbon nitride/wood." *Separation and Purification Technology* 279 (2021), 119788. doi.org/10.1016/j.seppur.2021.119788.

[53] Shah, H. U. Ahmad, K. Naseem, H. A. Parveena, S. A. M. and Rauf, A. "Water stable graphene oxide metal-organic frameworks composite (ZIF-67@ GO) for efficient removal of malachite green from water." *Food and Chemical Toxicology* 15, no. 4 (2021), 112312. doi.org/10.1016/j.fct.2021.112312.

[54] Adel, M. Ahmed, M. A. and Mohamed, A. A. "A facile and rapid removal of cationic dyes using hierarchically porous reduced graphene oxide decorated with manganese ferrite." *Flat Chem* 26 (2021), 100233. doi.org/10.1016/j.flatc.2021.100233.

[55] Khan, Z. U. Khan, W. U. Bakhtar, U. Ali, W. Ahmad, B. Ali, W. and Yap, P-S. "Graphene oxide/PVC composite papers functionalized with p-Phenylenediamine as high-performance sorbent for the removal of heavy metal ions." *Journal of*

Environmental Chemical Engineering 9, no. 5 (2021), 105916. doi.org/10.1016/j.jece.2021.105916.

[56] Li, Y. Dong, X. and Zhao, L. "Application of magnetic chitosan nanocomposites modified by graphene oxide and polyethyleneimine for removal of toxic heavy metals and dyes from water." *International Journal of Biological Macromolecules* 192 (2021), 118–125. doi.org/10.1016/j.ijbiomac.2021.09.202.

[57] Lee, S. Lingamdinne, L. P. Yang, J-K. Koduru, J. R. Chang, Y-Y. and Mu. N. "Biopolymer mixture-entrapped modified graphene oxide for sustainable treatment of heavy metal contaminated real surface water." *Journal of Water Process Engineering* 46 (2021), 102631. doi.org/10.1016/j.jwpe.2022.102631.

[58] Lee, S. Lingamdinne, L. P. Yang, J-K. Koduru, J. R. and Chang, Y-Y. "Potential electromagnetic column treatment of heavy metal contaminated water using porous Gd_2O_3-doped graphene oxide nanocomposite: characterization and surface interaction mechanisms." *Journal of Water Process Engineering* 41 (2021), 102083. doi.org/10.1016/j.jwpe.2021.102083.

[59] Khan, Z. U. Khan, W. U. Bakhtar, U. A. Wajid, A. Bilal, A. W. and Yap, P-S. "Graphene oxide modified thiosemicarbazide nanocomposite as an effective eliminator for heavy metal ions." *Journal of Environmental Chemical Engineering* 9, no. 5 (2021), 105916. doi.org/10.1016/j.jece.2021.105916.

[60] Chaabane, L. Beyou, E. and Baouab, M. H. V. "Preparation of a novel zwitterionic graphene oxide-based adsorbent to remove of heavy metal ions from water: modeling and comparative studies." *Advanced Powder Technology* 32, no. 7 (2021), 2502–2516. doi.org/10.1016/j.ijbiomac.2021.09.202.

[61] Archana, S. Jayanna, B. K. Ananda, A. Shilpa, B. M. Pandiarajan, D. Muralidhara, H. B. and Kumar, K. Y. "Synthesis of nickel oxide grafted graphene oxide nanocomposites—systematic research on chemisorption of heavy metal ions and its antibacterial activity." *Environmental Nanotechnology, Monitoring & Management* 16 (2021), 100486. doi.org/10.1016/j.enmm.2021.100486.

[62] Ibrahim, Y. Wadi, V. S. Ouda, M. Naddeo, V. Banat, F. and Hasan, S. W. "Highly selective heavy metal ions membranes combining sulfonated polyethersulfone and self-assembled manganese oxide nanosheets on positively functionalized graphene oxide nanosheets." *Chemical Engineering Journal* 428 (2022), 131267. doi.org/10.1016/j.cej.2021.131267.

[63] Abbasi, M. Safari, E. Baghdadi, M. and Janmohammadi, M. "Enhanced adsorption of heavy metals in groundwater using sand columns enriched with graphene oxide: lab-scale experiments and process modeling." *Journal of Water Process Engineering* 40 (2021), 101961. doi.org/10.1016/j.jwpe.2021.101961.

[64] Zarenezhad, M: Zarei, M. Ebratkhahan, M. and Hosseinzadeh, M. "Synthesis and study of functionalized magnetic graphene oxide for Pb^{2+} removal from waste water." *Environmental Technology & Innovation*, 22 (2021), 101384. doi.org/10.1016/j.eti.2021.101384.

[65] Naeimi, A. Amini, M. and Okati, N. "Removal of heavy metals from waste waters using an effective and natural bionanopolymer based on Schiff base chitosan/graphene oxide." *International Journal of Environmental Science and Technology* 19, no. 3 (2022), 1301. doi.org/10.1007/s13762-021-03247-9.

[66] Verma, M. Lee, I. Oh, J. Kumar, V. and Kim, H. "Synthesis of EDTA-functionalized graphene oxide-chitosan nanocomposite for simultaneous removal of inorganic

and organic pollutants from complex waste water." *Chemosphere* 287 (2022), 132385. doi.org/10.1016/j.chemosphere.2021.132385.
[67] Chen, R. Cheng, Y. Wang, P. Wang, Q. Wan, S. Huang, S. Su, R. Song, Y. and Wang, Y. "Enhanced removal of Co (II) and Ni (II) from high-salinity aqueous solution using reductive self-assembly of three-dimensional magnetic fungal hyphal/graphene oxide nanofibers." *Science of The Total Environment* 756 (2021), 143871. doi.org/10.1016/j.scitotenv.2020.143871.
[68] Farahani, F. V. Amini, M. H. A. Ahmadi, S. H. and Zakaria, S. A. "Investigation of layered double hydroxide/carbon dot nanocomposite on removal efficiency of Pb^{2+} from aqueous solution." *Journal of Molecular Liquids* 338 (2021), 116774. doi.org/10.1016/j.molliq.2021.116774.
[69] Wang, Y. Ding, G. Lin, K. Liu, Y. Deng, X. and Li, Q. "Facile one-pot synthesis of ultrathin carbon layer encapsulated magnetite nanoparticle and graphene oxide nanocomposite for efficient removal of metal ions." *Separation and Purification Technology* 266 (2021), 118550. doi.org/10.1016/j.seppur.2021.118550.

3 Metal-Organic Frameworks (MOFs) for Water Treatment

Manjula Nair[1*] and Kajal Panchal[2]
[1]Heriot Watt University, Knowledge Park, Dubai
[2]School of Chemical Sciences, Central University of Gujarat, Gandhinagar, Gujarat, India
*Corresponding author: Email: manjulanair13@yahoo.com

CONTENTS

3.1 Introduction 45
    3.1.1 The Chemistry of MOFs 46
    3.1.2 Types of MOF 46
3.2 Water Contamination...47
    3.2.1 The use of MOFs for Water Remediation 48
    3.2.2 The Removal of Inorganic Pollutants...48
    3.2.3 Removal of Organic Pollutants 50
    3.2.4 The Removal of Dyes...51
3.3 MOF Based Materials for Water Treatment 52
    3.3.1 MOF Based Aerogel/Hydrogel...53
    3.3.2 MOF Derived Carbon Material 54
    3.3.3 Hydrophobic MOFs and Their Composites 55
    3.3.4 Magnetic Framework Composites..56
3.4 Conclusion...56
References...56

3.1 INTRODUCTION

MOF membranes have found widespread use in water treatment, water filtration, and dye degradation. MOFs are spongy substances, having arrays of positively charged metal ions encircled by organic linkers. The organic-inorganic hybrid frameworks demonstrate a hollow cage-like architecture, due to which, the MOFs have a high intrinsic surface area and many adsorption sites, as shown in Figure 3.1. MOFs have the advantage that they can be characterized for specific applications by altering properties such as porosity, stability, and particle morphology [1].

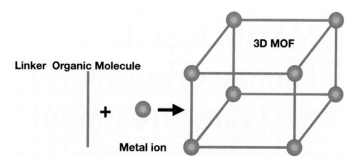

FIGURE 3.1 Schematic of the construction of a 3D MOF.

This discussion of MOFs excludes zeolites, which, although crystalline and microporous, do not possess synthetic flexibility.

3.1.1 The Chemistry of MOFs

A MOF comprises of an organic-inorganic composite skeleton and acts as a coordination polymer. The organic units (linkers) are mono, di, tri, or tetravalent ligands. Some ligands that are used are dicarboxylic and tricarboxylic acids, amines, and imidazoles. The secondary building units (SBU) are metal ion clusters. The inorganic nodes of the MOFs can be synthesized from monovalent, divalent, trivalent, or tetravalent metal cations. Group 4 metal-based MOFs have gained particular attention as they provide better chemical stability and structural tunability [2].

3.1.2 Types of MOF

There are about 90,000 MOF structures that have been synthesized depending on their end use. MOFs find application in water treatment, gas storage, catalysis, and solid state electrolytes. There are MOF-based composites in which a MOF is combined with other materials such as nanoparticles, quantum dots, natural enzymes, and polymers (Figure 3.2). The composites have better properties in comparison with single-component materials [4]. Post synthetic modification can also be used to upgrade the efficiency of MOF materials.

Primarily, MOFs are classified as:

- **Normal MOFs** (First generation)
- **Structural MOFs** (Second generation) which are synthesized by chemically modifying the first-generation MOFs.
- **Smart MOFs** (Third generation) which wave synthesized by the integration of biomolecules, such as some organic drugs, cations, toxins, gases, and certain bioactive molecules.

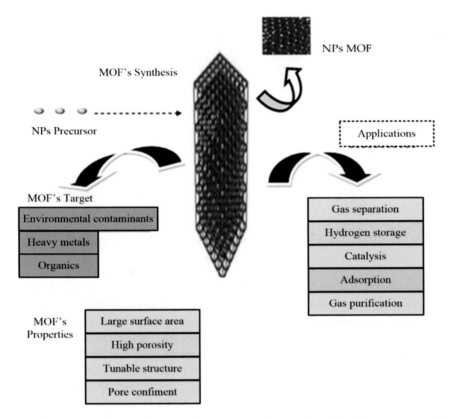

FIGURE 3.2 Properties and uses of MOFs ([3]). License: http://creativecommons.org/licenses/by/4.0/).

MOFs are classified as rigid or flexible, based on their framework structure. They are also classified as crystalline or amorphous. Crystalline MOFs have a well-defined arrangement with the metal nodes connected by organic linkers. Amorphous MOFs lack long-range order and are generally obtained from the parent crystalline material's structural collapse or melt quenching [5].

3.2 WATER CONTAMINATION

Anthropogenic activities have contributed to the decline in water quality. Growing population, industrialization, mining, and pollution because of industrial effluents and domestic sewage have all contributed to a deterioration in water quality. Some of these activities have resulted in nutrient loading and introducing heavy metal ions such as mercury in the water [6].

According to their chemical compositions, contaminants in water are predominantly classified into organic and inorganic pollutants. Many methods, like coagulation, flocculation, precipitation, and membrane technology, have been employed for water remediation. Membrane-based processes are expensive and energy-intensive as compared to other conventional methods. Combining two or more processes, hybrid methods have also been employed to treat water.

3.2.1 The use of MOFs for Water Remediation

There is ongoing research on using MOFs as adsorbents of inorganic pollutants. In contrast to the traditional adsorbents, MOFs exhibit high specific surface area, fine pore size distribution, large pore volume, and remarkably long-lasting porosity. Though many MOFs, namely MOF-5 and MIL-101-V, face deterioration during the dislocation of ligands or hydrolysis when coming in contact with water, some MOFs are stable in water and can adsorb contaminants. Incorporating water repellent groups can also enhance the stability of MOFs [7].

Various MOFs, for example defective MOFs, hierarchal-pore MOFs, and MOF-based hybrids, have been used for waste water treatment [8].

- Defective MOFs: Imperfections can be introduced into MOFs to modify their properties. This provides additional active sites and improves diffusion and mass transfer by creating more pore spaces. This constructional imperfection in MOFs has been advantageous in catalysis, adsorption, purification, and so forth. [9].
- Hierarchal-pore MOFs: Stratified pores incorporate multi-domain or multi-level pore apportionments within a porous system. MOFs with multiple porosities, micropores, macropores, and mesopores, have wider applications. For example, mesopores can facilitate the movement of large molecules, while micropores are size-selective. The integration of porosity allows for better substrate transportation [10].
- MOF-based composites: The integration of magnetic NPs and MOFs has been proven to be effective for pollutant sequestration [11]. MOFs are robust, and several active species can be encapsulated without affecting the stability of the original framework [12].

3.2.2 The Removal of Inorganic Pollutants

Common inorganic pollutants include arsenic, nitrates, fluoride, heavy metals, and radioactive materials. Mining activities, metal refining, and mineral extraction pollute the water with heavy metals such as, Cd, Cr, Cu, Hg, Pb, and Zn. Natural sources such as volcanoes, soil erosion, and the disintegration of rocks also add to the problem [13]. Ion exchange processes using cation exchangers have eliminated heavy metal ions in water. However, the removal of heavy metals depends on factors such as pH, temperature, and contact time [14]. Zeolites have also been employed because of their low cost and abundance. However, it was found that the rate of adsorption reduces with time because of the decline in the number of adsorption sites [15].

Metal nodes, organic linkers, and moieties integrated on the surfaces of MOFs could chelate heavy metal ions to form stable complexes and have been found efficient to adsorb ions like Cu(II), Pb(II), Zn(II), U(VI), Pd(II), Pt(IV), and Au(III) [16].

Wang et al. demonstrated the adsorption process of Pb(II) in six substituted MOFs: MIL-53(Fe), MIL-101(Fe), UIO-66(Zr), IRMOF-3(Zn), MOF-5(Zn), and ZIF-8(Zn) [17]. The adsorption kinetics and isotherms showed that Zinc containing MOFs exhibited advanced adsorption efficiency. The removal of Pb by this method introduced Zn into the environment, which is also a contaminant, albeit less harmful. PSM on UiO-66-NH_2(Zr) with thiourea, isothiocyanate, and moieties have been inspected to eliminate toxic heavy metal ions, namely Pb(II), Cr(III), Cd(II) and Hg(II) from waste water [18]. The sulfur containing structures possess effective integration with the metal ions.

ED-MIL-101 has been found to be a good adsorbent for Pb(II) as compared to MIL-101 due to the good coordination between Pb(II) and EDTA [18]. NH_2-MIL-53, with a nitrogen-based functionality, has a higher Pb(II) ion adsorption efficiency, which can be tuned by changing the concentration of the amino group in the material [19].

Wang et al. observed that the $Cu_3(BTC)_2$-SO_3H skeleton is efficient in removing Cd(II). This was prepared by PSM with sulfonic acid on $Cu_3(BTC)_2$. Adsorption worked best at a pH of around 6. At pH values less than 6, the protons occupied the active sites, and at pH values greater than 6, Cd ions were precipitated as hydroxides. The adsorption process shows chelation between Cd ions and sulfonic groups of MOFs [20].

TMU-5 can adsorb Cd(II), Co(II), Cr(III), Cu(II), and Pb(II) in waste water showing good adsorption performance, which is attributed to its highly ordered 3D structure of pores lined with azine moieties [21].

Recently, experimentation has been conducted on utilizing modified magnetic MOF composites for heavy metal ion removal; these are more suitable for industrial applications. Post synthetic modification (PSM) of a magnetic MOF hybrid containing Fe_3O_4@SiO_2 MNPs and UiO-66(Zr), with -SH moiety, has yielded a composite with increased Hg^{+2} adsorbent elimination [22]. The adsorption is fast, and the interference of additional metal ions does not affect the performance of this composite. It can also be revitalized and reutilized up to 5 times with no reduction in efficiency.

In another study by Alqadami et al. Fe_3O_4@AMCA -MIL53 (Al) was employed for the uptake of U(VII) and Th(IV) metal ions from an aqueous environment. The adsorption capacity was good, and the addition of 0.01M HCl could easily desorb adsorbed metals.

Radionuclides can be adsorbed using defects in the MOF framework. Yien et al. integrated dodecanoic acid in manufacturing hierarchal porous UiO-66 (HP-UiO-66), which showed rapid adsorption of U [23]. This framework was also efficient in eliminating and reutilizing arsenic and Pt^{4+} using CH_3COOH and TFAA as modulators.

3.2.3 Removal of Organic Pollutants

Organic water contamination includes detergents, disinfection by-products, food processing waste, insecticides and herbicides, petroleum hydrocarbons and lubricants and fuel combustion by-products. Phosphates, nitrate, sulfate, ammonium nitrate, and nitrites are also considered organic pollutants. There is a group of compounds, called emerging organic contaminants (EOC), including veterinary products, pharmaceutical products, personal care products, and engineered nanomaterials. Currently used technologies are ineffective in eliminating these contaminants [24]. The use of personal care products that form part of this group is increasing, and the lack of decontamination protocols has led to their increasing concentrations in water [25]. Some organic pollutants can be broken down easily; however, a class of compounds called persistent organic pollutants (POPs) pose long-term risks. These are DDT, Endrin, Aldrin, and dioxins [26]. Some of these toxic compounds enter the food chain and can negatively affect human health.

Organic pollutants can be removed by filtration, coagulation [27,28], chemical precipitation, advanced oxidation processes, or adsorption [29]. Adsorption is the most used technique owing to its simplicity, low energy requirement, and the fact that it does not require any pre-treatment.

MOFs find application in eliminating organic waste due to their versatility and the availability of open metal sites. Since these open metal sites can be Lewis acid sites, they can selectively remove Lewis bases via a coordinate bond formation [30].

Recently, functional MOFs have been used to remove EOCs. MIL-101-OH exhibits better adsorption capacity than MIL-101 for the removal of organics like Bisphenol A, Naproxen, ketoprofen, and the like. The adsorption capacity is enhanced linearly with the enhancement at several H adsorption sites, confirming hydrogen bonding [31]. Functionalized UiO-66s were found to be very productive in the adsorption of diclofenac sodium because of favorable ionic association.

BUT-12 showed good uptake capacity for nitrofurantoin and nitrofurazone. BUT-13 exhibited fast and remarkable adsorption for nitrofurantoin, nitrofurazone, ornidazole, and chloramphenicol. This is attributed to the hydrophobic pore surfaces, and appropriate pore sizes of both these materials, and also to their good recyclability [32].

A class of antibiotics, called nitroimidazole antibiotics (NIABs) used to treat anaerobic and protozoan bacterial infections are also found in hospital effluents [33]. They are highly soluble in water and have low biodegradability. This class of pollutants was efficiently adsorbed by urea-MIL-101. The adsorption mechanism suggests hydrogen bonding between the -NO_2 of the NIABs and the -NH_2 of the MOF. The -NH_2 is inaugurated into the MOF via urea or melamine grafting.

The use of pesticides in agriculture has resulted in residues that pose health hazards. Different MOFs such as MIL-101(Cr), MIL-53(Cr), UiO-66(Zr), UiO-67(Zr), ZIF-8(Zn), ZIF-67(Co), and Cu-BTC were employed as adsorbents for eliminating pesticides from water. MIL-53(Cr) was found to be effective in the eradication of 2,4-dichlorophenoxyacetic acid [34]. MOF-235(Fe) was a good adsorbent for bentazon, clopyralid, and isoproturon from water [35]. Though

MOF-235(Fe) was a superior adsorbent to zeolites, it was not very stable in the aqueous phase, and its reusability was poor. Cu-BTC, a Cu-based MOF, was found to be efficient in removing 14C-ethion, a carcinogenic and toxic pesticide. The $_{14}$C-ethion formed a coordinate bond with the Cu (II) of the MOF [36].

Tetracycline hydrochloride is an antibiotic employed for treating bacterial infections and a feed additive in animal husbandry. Traces of this antibiotic are found in meat, milk, and fish and long-term intake can cause health problems such as gastrointestinal disorders, hepatoxicity, and antibiotic resistance [37]. MOF-1 was found to be a good absorbent for removing tetracycline in milk. The MOF-1 interacts with tetracycline through π–π interactions, hydrogen bonding, and electrostatic interactions [38].

3.2.4 THE REMOVAL OF DYES

Dyes released by industry pollute water bodies and can negatively affect human health, especially synthetic dyes like methylene blue, methyl orange, Congo red, and the like. Chromium-based dyes have been found to be carcinogenic, anthroquinone based dyes are persistent, while water-soluble dyes produce highly colored effluents. The dyes from the textile industry prevent penetration of light, which affects photosynthesis in plants [39].

Conventional coagulation, membrane separation, and oxidative processes require expensive and sophisticated technology [40]. Adsorption techniques are far superior in removing dyes, especially on a large scale, because of their easy handling and high functioning. MOFs and MOF composites have been effective in separating dyes because of the availability of large pore sizes, the functionalization of organic linkers, and defect engineering [41].

The elimination of contaminants from water uses the adsorption phenomenon. Adsorption is entirely a surface property in which adsorbate is integrated with adsorbent through the formation of different bonds or interactions: ionic bonding, H-bonding, acid-base linkage, π–π linkage, hydrophobic linkage, and pore selective adsorption.

MOF-based material removes contaminants from water resources and shows reusability and no toxicity, as shown in Figure 3.3. Fe incorporated MOFs have been found to be efficient because the magnetic properties acquired by the MOFs make separation easier after adsorption [42]. Factors such as pH, temperature, surface area, and pore volume influence the adsorption of dyes on MOFs. Water stable MOFs such as zirconium-based MOFs and UiO-66 have been found to be productive in the uptake of cationic and anionic dyes [43]. UiO-66 showed better adsorption with anionic dye (methyl orange) under acidic and neutral conditions than did MB, a cationic dye, where adsorption is better under strongly basic conditions. This makes UiO-66 effective in the segregation of anionic dyes from cationic dyes. Temperature alteration also affects adsorption, for example, the adsorption of methyl orange decreases with an increase in temperature, as it is an endothermic process. The benzene rings of methyl orange and methylene blue

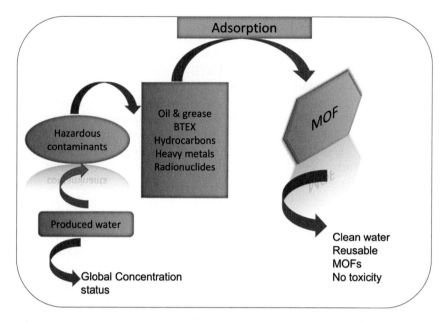

FIGURE 3.3 MOF applications to eliminate pollutants ([3]). License: http://creativecommons.org/licenses/by/4.0/

exhibit a π–π stacking bonding with the benzene rings of UiO-66, due to which they were adsorbed onto the MOF.

Recently, GO (graphic oxide) has been incorporated into MOFs. These have a high surface area and can undergo surface modification with carboxyl, hydroxyl, and carbonyl groups that can then be attached to MOF surfaces for suitable applications. Mn-UiO-66@GO-NH$_2$ has shown excellent adsorption of Congo red, the possible mechanism for this being a combination of electrostatic attraction, π-π stacking, and hydrogen bonding [44].

The structural moiety of the organic ligands influences the adsorption of dyes on MOFs. An experiment conducted by Nanthamathee et al. showed that UiO-66-X (X = H, NH$_2$, NO$_2$), and UiO-66-NH$_2$ showed extreme adsorption capacity due to the creation of more intermolecular forces [45]. With neutral dyes like phenol red, the above-mentioned adsorption capacity of the MOFs reduces due to a lack of electrostatic interaction. The steric hindrance of the phenol red molecule also affects the π-π interaction.

The surface modification of MOFs enhances dye removal, although the transformation process is costly and sluggish.

3.3 MOF BASED MATERIALS FOR WATER TREATMENT

MOFs, being flexible and versatile in adapting to various modifications for specific applications, is widely used in water treatment. The limitations of MOF

Metal-Organic Frameworks

are lack of stability [46] and its inability to be separated in solid form. However, precise modifications have been made in structure and shape by converting the powder form into spheres [47], membranes [48], chains [49], and aerogels or hydrogels [50] like elevated dimensional structures. The engineered structures have restricted the utilization of conventional MOFs. This can improve the traditional MOFs natural instability and low conductivity by combining them with a highly stable and conductive material like graphene [51]. Thus, enhancing their properties has opened the door for the MOFs to be used in water remediation. The advanced forms of MOFs can be classified according to the following features:

- MOFs with enhanced extrinsic properties
- MOFs that have enhanced intrinsic properties such as more stability, greater wettability, larger surface areas, and a quicker reaction to external stimuli compared to the traditional MOFs.

Based on the these qualities, four different categories of MOF have advanced the water treatment process.

3.3.1 MOF Based Aerogel/Hydrogel

The practical use of the powder form of MOFs and their modification into desirable shapes is always a difficult task. It is essential to convert them into macroscopic shapes, to facilitate easy handling during practical application. The molded shape should be such that it preserves its porous structure and its diffusion kinetics [52]. The recent trends in MOF engineering are to shape MOFs into aerogels/ hydrogels to bring convenience and to provide better service. There are in-situ and ex-situ synthesis methods for MOF-based aerogels/hydrogels. In situ growth of MOF crystals into the aerogel is a difficult task and faces the following problems during synthesis.

- The in-situ growth of MOFs brings inflexibility, which somehow impedes their modification into desirable shapes [53].
- There is insufficient dispersion and an incompatibility of the MOF-matrix structure due to the aggregation of MOF particles.
- The synthesis procedure to get gelation modifies processing and creates tedious experimental conditions such as the use of various solvents and the maintenance of pH, time, and temperature [54].

Defining these challenges and modifying MOFs to be used in water treatment is a new focus in research. MOF-based aerogels are formed by polysaccharides, namely, agar, cellulose, and alginates [55,56]. These polysaccharides easily form cross-links, provide a gel-like structure, and have good availability. Zhang et al. synthesized a zeolitic imidazole framework (ZIF) MOF by utilizing cheap and easily available agar polysaccharides for water decontamination [57]. However, its poor reusability hinder its utilization as an adsorbent in water treatment. So, Zhang and co-workers have made efforts to enhance their accessibility for dye removal by integrating ZIF particles into substrate matrices. Carbon nitride was selected to serve as a substrate matrix due to its layered nanosheet structure. C_3N_4 enhances the accumulation or aggregation of particles and inhibits extra growth of the MOF particles. Therefore, Zhang and co-workers incorporated aerogel into carbon nitride dispersed with ZIF-8 particles. This ZIF-8/C_3N_4 aerogel has shown its efficiency in removing Congo red dye. MOF-based aerogels fabricated with cellulose found good use in effectively eliminating toxic ions from water. Cellulose aerogel served as a sorbent but had some disadvantages, such as low physical and mechanical stability and fragility. To improve its stability and enhance its applicability, Bo et al. fabricated ZIF-8- cellulose hybrid aerogel and assessed in usability for the removal of Cr(VI) ions [58]. The starting concentration of 20 ppm has been reduced and effectively eliminates 90.79% of Cr(VI) ions.

By the same token, polysaccharide alginate shows its application in matrices for synthesizing and modifying MOF particles. Alginate is a natural polymer and can be allowed to form ionotropic gelation and create hydrogels [59]. The disadvantage of in-situ growth, where aggregation of MOF particles restricts its efficiency, has been mitigated by improving its preparation method. Zhuang et al. synthesized a ZIF-67 MOF, where alginate worked as matrices [60]. The as-synthesized ZIF-67 MOF demonstrated its efficiency in the removal of tetracycline antibiotics. MOF-based composites have also been synthesized using reduced graphene oxide as a base or support matrix and can be depicted as MOF/rGA [61]. This hybrid material finds very successful application in water decontamination due to the robust nature of an integrated rGO.

Along with this, it provides plentiful functional group sites for integration and offers ease in shape modification. Mao et al. synthesized reduced graphene oxide aerogel (rGA) incorporated ZIF-8 and proved its effective application to serve as an adsorbent. ZIF-8/rGA is used in removing toxic heavy metal ions and as an adsorbent for organic liquids [62].

3.3.2 MOF Derived Carbon Material

Nanoporous carbon material possesses a high surface area and enhanced stability and is of value in water treatment. However, its wide pore size distribution and its disordered framework limits its application. MOFs attract attention for the synthesizing of carbons when they have an ordered pore size range. The MOF serves as a template to obtain porous carbonaceous material. Along with this, the MOF also introduces porosity into the carbon framework. Thus, the incorporation of metal ions occurs effectively. So, MOF-derived carbon (MDC) and its hybrid structures are extensively used in water decontamination. ZIF-67 MOF serves as a template for the formation of porous carbon material. Torad and co-workers observed its application in eliminating methylene blue and offering sites for the adsorption of metal ions [63]. Liu et al. observed that the ZIF-67 MOF works as a template and synthesizes an effective adsorbent as carbon with effective magnetic properties. This magnetic carbon has been used to draw out herbicide, namely phenyl urea, from grapes and the bitter gourd [64]. Zinc-based MDC possesses unique properties as the zinc metal provides a low boiling point range and is used to derive the porous carbon material. Bhadra and co-workers use a biomass, namely Bio-MOF-1, which is synthesized using a skeleton containing Zn, has biphenyl dicarboxylate as a linker. The complete framework is engulfed by adenine biomolecules. Bisphenol has been removed using bio-MOF-1. The available porous carbon has less adsorption capacity than the Bio-MOF-1, in other words, it provides five times better adsorption and elimination of bisphenol [65]. Such enhanced uptake of contaminants is due to the hydrogen bonding provided by oxygen and nitrogen-containing functional groups.

3.3.3 Hydrophobic MOFs and Their Composites

The oil-water separation has been achieved by utilizing the tunable wettability of MOFs. However, water decontamination is inhibited due to the degradation of MOFs in hydrophilic media, which affects its application in water treatment. Therefore, the stability of MOFs is enhanced by building them waterproof and by successfully improving their wettability. Hydrophobic MOFs and their composites are focused on applications in oil-water separation. Conventional methods employed in oil-water separation limit their future utilization because of their low function ability and require high maintenance and high-power supply input [66]. The oil-water separation is a phenomenon that is entirely related to the surface wettability and is improved by engineering the hybrid materials. Surface wettability is a property that analyzes the functioning of a solid surface. A material's high hydrophobic property exhibits oil removal, and a high hydrophilic character exhibits the property of removing water. So, the development of some advanced materials is required to improve the functioning of the surface wettability of the material. The engineered hydrophobic MOFs provide less adhesion toward water, remove oil, and decontaminate the water. There are various strategies to synthesize hydrophobic MOFs where hydrophobic linkers, hydrophobic polymer layers, and some other hydrophobic materials provide advanced features and help in oil-water separation [67].

Hydrophobic MOFs provide porous structures with water repellent properties; they are prioritized against conventional hydrophobic material. Roy and co-workers have used alkyl functionalizing linkers and synthesized NMOF-1, which is highly hydrophobic [68]. They observed its utility in cleaning oil spills and hence it is a good candidate for water treatment. Zhang et al. synthesized a hydrophobic MOF, named UPC-21. The water repellent property is provided by the pentiptycene ligand and exhibits a zero-degree angle with oil. The water repellent UPC-21 MOF eliminates and segregates hexane, toluene, crude oil, naphtha, and gasoline from water sources [69]. The as-synthesized MOF exhibits good viscosity and therefore indicates its high efficiency towards pollutant and oil separation.

3.3.4 MAGNETIC FRAMEWORK COMPOSITES

Engineered magnetic framework composites (MFC) with improved physical properties and a response towards external magnetic stimulation, appears to be attractive for use in water treatment. Solid-phase segregation of adsorbents using sorbents in water treatment for a large volume of water needs easy segregation processes [70]. Conventionally available techniques are limited in their applications because of laborious operating procedures, particularly filtration and centrifugation at high speed. Therefore, a simple segregation of solid-phase pollutants and rapid mobility of mass transfer is achieved by synthesizing and using MFC [71]. This advancement in the MOF structure is required, since it exhibits less magnetic properties. Advancement could be introduced by incorporating magnetic nanoparticles. The synthesis techniques and MOF play a major role in the water decontamination application. The techniques used in MFC synthesis are: 1) core-shell skeleton of functionalized MNP integrating the MOF; 2) magnetic core component synthesis by metal ions serving as a template for the MOF; 3) the incorporation of MNPs into the skeleton of the MOF. Xiao and co-workers synthesized MFC NiCo/Fe_3O_4-MOF-74 using tri-metallic MOF-74. Initially, NiCoFe-MOF-74 undergoes some heat treatment to exhibit better sorption efficiency for the effective removal of tetracycline from the waste water [72]. Huang and co-workers synthesized Zirconium-containing MFC through a layer-by-layer approach. Zr containing UiO-66 is prepared by a ligand exchange process using mercaptoacetic acid. Initially, a $Fe_3O_4@SiO_2$ core is grown and integrated with MAA functionalized MOF and this exhibits good removal of mercury ions from contaminated water [73].

3.4 CONCLUSION

Although there has been considerable advancement in MOF synthesis, there are several challenges in their stability and reusability. Most of the published research primarily covers a maximum usage of 10 cycles. The reusability of MOFs matters in their deployment on a commercial scale. Research must be directed at synthesizing MOF composites with desirable properties to scale up their application.

REFERENCES

[1] Baumann, A. E. Burns, D. A. Liu, B. and Sara Thoi, V. "Metal-organic framework functionalization and design strategies for advanced electrochemical energy storage devices." *Communications Chemistry* 2, no. 1 (2019), 1–14. doi: 10.1038/s42004-019-0184-6.

[2] Yuan, S. Qin, J-S. Lollar, C. T. and Zhou, H-C. "Stable metal–organic frameworks with group 4 metals: current status and trends." *ACS Central Science* 4, no. 4 (2018), 440–450. doi: 10.1021/acscentsci.8b00073.

[3] Gul Zaman, H. Baloo, L. Pendyala, R. Singa, P. K. Ilyas, S. U. and Mohamed Kutty, S. R. "Produced water treatment with conventional adsorbents and MOF as an alternative: a review." *Materials* 14, no. 24 (2021), 7607. doi: 10.3390/ma14247607.

[4] Chen, L. Zhang, X. Cheng, X. Xie, Z. Kuang, Q. and Zheng, L. "The function of metal–organic frameworks in the application of MOF-based composites." *Nanoscale Advances* 2, no. 7 (2020), 2628–2647 doi: https://doi.org/10.1039/D0NA00184H

[5] Bennett, T. D. and Cheetham, A. K. "Amorphous Metal–Organic Frameworks." *Accounts of Chemical Research* (2014),47 1555–1562.

[6] Vareda, J. P. Valente, A. J. M. and Durães, L. "Assessment of heavy metal pollution from anthropogenic activities and remediation strategies: A review." *Journal of Environmental Management* 246 (2019), 101–118. doi: https://doi.org/10.1016/j.jenvman.2019.05.126.

[7] Wu, T. Shen, L. Luebbers, M. Hu, C. Chen, Q. Ni, Z. andMasel, R. I. "Enhancing the stability of metal–organic frameworks in humid air by incorporating water repellent functional groups." *Chemical Communications* 46, no. 33 (2010), 6120–6122. doi: https://doi.org/10.1039/C0CC01170C.

[8] Liu, X. Shan, Y. Zhang, S. Kong, Q. andPang, H. "Application of metal organic framework in waste water treatment." *Green Energy & Environment* (2022), doi: https://doi.org/10.1016/j.gee.2022.03.005.

[9] Xiang, W. Zhang, Y. Chen, Y. Liu, C-J. and Tu, X. "Synthesis, characterization and application of defective metal–organic frameworks: current status and perspectives." *Journal of Materials Chemistry A* 8, no. 41 (2020), 21526–21546. doi: https://doi.org/10.1039/D0TA08009H.

[10] Feng, L. Wang, K-Y. Lv, X-L. Yan, T-H. and Zhou, H-C. "Hierarchically porous metal–organic frameworks: synthetic strategies and applications." *National Science Review* 7, no. 11 (2020), 1743–1758. doi: https://doi.org/10.1093/nsr/nwz170.

[11] Alqadami, A. Naushad, M Alothman, Z. and Ghfar, A. "Novel metal–organic framework (MOF) based composite material for the sequestration of U (VI) and Th (IV) metal ions from aqueous environment." *ACS Applied Materials & Interfaces* 9, no. 41 (2017), 36026–36037. doi: https://doi.org/10.1021/acsami.7b10768.

[12] Juan-Alcaniz, J. Gascon, J. and Kapteijn, F. "Metal–organic frameworks as scaffolds for the encapsulation of active species: state of the art and future perspectives." *Journal of Materials Chemistry* 22, no. 20 (2012), 10102–10118. doi: https://doi.org/10.1039/C2JM15563J.

[13] Srivastav, A. L. and Ranjan, M. "Inorganic water pollutants." *Inorganic Pollutants in Water*, (2020), 1-15. doi:10.1016/b978-0-12-818965-8.00001-9.

[14] Gode, F. and Pehlivan, E. "Removal of chromium (III) from aqueous solutions using Lewatit S 100: the effect of pH, time, metal concentration and temperature." *Journal of Hazardous Materials* 136, no. 2 (2006), 330–337. doi: https://doi.org/10.1016/j.jhazmat.2005.12.021.

[15] Elboughdiri, N. and Garcia, H. A. "The use of natural zeolite to remove heavy metals Cu (II), Pb (II) and Cd (II), from industrial waste water." *Cogent Engineering* 7, no.1 (2020), 1-13 doi: https://doi.org/10.1080/23311916.2020.1782623.

[16] Ghaedi, A. M. Panahimehr, M. Shirazi Nejad, A. R. Hosseini, S. J. Vafaei, A. and Baneshi, M. M. "Factorial experimental design for the optimization of highly selective adsorption removal of lead and copper ions using metal organic

framework MOF-2 (Cd)." *Journal of Molecular Liquids* 272 (2018), 15–26. doi: https://doi.org/10.1016/j.molliq.2018.09.051.

[17] Tahmasebi, E. Masoomi, M. Y. Yamini, Y. and Morsali, A. "Application of mechanosynthesized azine-decorated zinc (II) metal–organic frameworks for highly efficient removal and extraction of some heavy-metal ions from aqueous samples: a comparative study." *Inorganic Chemistry* 54, no. 2 (2015), 425–433. doi: https://doi.org/10.1021/ic5015384.

[18] Luo, X. Ding, L. and Luo, J. "Adsorptive removal of Pb (II) ions from aqueous samples with amino-functionalization of metal–organic frameworks MIL-101 (Cr)." *Journal of Chemical & Engineering Data* 60, no. 6 (2015), 1732–1743. doi: https://doi.org/10.1021/je501115m.

[19] Ricco, R. Konstas, K. Styles, M. J. Richardson, J. J. Babarao, R. Suzuki, K. Scopece, P. and Falcaro, P. "Lead (II) uptake by aluminium based magnetic framework composites (MFCs) in water." *Journal of Materials Chemistry A* 3, no. 39 (2015), 19822–19831. doi: https://doi.org/10.1039/C5TA04154F.

[20] Kobielska, P. A. Howarth, A. J. Farha, O. K. and Nayak, S. "Metal–organic frameworks for heavy metal removal from water." *Coordination Chemistry Reviews* 358 (2018), 92–107. doi: https://doi.org/10.1016/j.ccr.2017.12.010.

[21] Rasheed, T. Hassan, A. A. Bilal, M. Hussain, T. and Rizwan, K. "Metal-organic frameworks-based adsorbents: a review from removal perspective of various environmental contaminants from waste water." *Chemosphere* 259 (2020, 127369. doi: https://doi.org/10.1016/j.chemosphere.2020.127369.

[22] Huang, L. He, M. Chen, B. and Hu, B. "A mercapto functionalized magnetic Zr-MOF by solvent-assisted ligand exchange for Hg^{2+} removal from water." *Journal of Materials Chemistry A* 4, no. 14 (2016), 5159–5166. doi: https://doi.org/10.1039/C6TA00343E.

[23] Yin, C. Liu, Q. Chen, R. Liu, J. Yu, J. Song, D. and Wang, J. "Defect-induced method for preparing hierarchical porous Zr–MOF materials for ultrafast and large-scale extraction of uranium from modified artificial seawater." *Industrial & Engineering Chemistry Research* 58, no. 3 (2018), 1159–1166. doi:10.1021/acs.iecr.8b04034.

[24] Lapworth, D. J. Baran, N. Stuart, M. E. and Ward, R. S. "Emerging organic contaminants in groundwater: a review of sources, fate and occurrence." *Environmental Pollution* 163 (2012), 287–303. doi: https://doi.org/10.1016/j.envpol.2011.12.034.

[25] Bu, Q. Wang, B. Huang, J. Deng, S. and Yu, G. "Pharmaceuticals and personal care products in the aquatic environment in China: a review." *Journal of Hazardous Materials* 262 (2013), 189–211. doi: https://doi.org/10.1016/j.jhazmat.2013.08.040.

[26] Hossain, M. Mosharraf, K. M. Islam, N. and Rahman, I. M. M. "An overview of the persistent organic pollutants in the freshwater system." *Ecological Water Quality-Water Treatment and Reuse* (2012), 455–496.

[27] Chen, L. Zhang, X. Cheng, X. Xie, Z. Kuang, Q. and Zheng, L. "The function of metal–organic frameworks in the application of MOF-based composites." *Nanoscale Advances* 2, no. 7 (2020), 2628–2647. doi: https://doi.org/10.1039/D0NA00184H.

[28] Chen, Z. Yang, B. Wen, Q. and Chen, C. "Evaluation of enhanced coagulation combined with densadeg-ultrafiltration process in treating secondary

effluent: organic micro-pollutants removal, genotoxicity reduction, and membrane fouling alleviation." *Journal of Hazardous Materials* 396 (2020), 122697. doi: https://doi.org/10.1016/j.jhazmat.2020.122697.

[29] Rashed, M. N. "Adsorption technique for the removal of organic pollutants from water and waste water." *Organic Pollutants-Monitoring, Risk and Treatment* 7 (2013), 167–194.

[30] Kokçam-Demir, U. Goldman, A. Leili Esrafili, M. G. Morsali, A. Weingart, O. and Janiak, C. "Coordinatively unsaturated metal sites (open metal sites) in metal–organic frameworks: design and applications." *Royal Society of Chemistry* (2020), 49, 2751–2798.

[31] Sarker, M. Song, J. Y. and Jhung, S. H. "Adsorptive removal of anti-inflammatory drugs from water using graphene oxide/metal-organic framework composites." *Chemical Engineering Journal* 335 (2018), 74–81. doi: https://doi.org/10.1016/j.cej.2017.10.138.

[32] Wang, L. Zhao, X. Zhang, J. and Xiong, Z. "Selective adsorption of Pb (II) over the zinc-based MOFs in aqueous solution-kinetics, isotherms, and the ion exchange mechanism." *Environmental Science and Pollution Research* 24, no. 16 (2017), 14198–14206. doi: 10.1007/s11356-017-9002-9.

[33] Rivera-Utrilla, J. Sánchez-Polo, M. Ferro-García, M. Á. Prados-Joya, G. and Ocampo-Pérez, R. "Pharmaceuticals as emerging contaminants and their removal from water. A review." *Chemosphere* 93, no. 7 (2013), 1268–1287. doi: https://doi.org/10.1016/j.chemosphere.2013.07.059.

[34] Jung, B. K. Hasan, Z. and Jhung, S. H. "Adsorptive removal of 2, 4-dichlorophenoxyacetic acid (2, 4-D) from water with a metal–organic framework." *Chemical engineering journal* 234 (2013), 99–105. doi: https://doi.org/10.1016/j.cej.2013.08.110.

[35] De Smedt, C. Spanoghe, P. Biswas, S. Leus, K. and Van Der Voort, P. "Comparison of different solid adsorbents for the removal of mobile pesticides from aqueous solutions." *Adsorption* 21, no. 3 (2015), 243–254. doi: 10.1007/s10450-015-9666-8.

[36] Abdelhameed, R. M. H. Abdel-Gawad, C. M. Silva, J. Rocha, B. Hegazi, and Silva, A. M. S. "Kinetic and equilibrium studies on the removal of 14C-ethion residues from waste water by copper-based metal–organic framework." *International Journal of Environmental Science and Technology* 15, no. 11 (2018), 2283–2294. doi: 10.1007/s13762-017-1624-4.

[37] Gajda, A. Nowacka-Kozak, E. Gbylik-Sikorska, M. and Posyniak, A. "Tetracycline antibiotics transfer from contaminated milk to dairy products and the effect of the skimming step and pasteurization process on residue concentrations." *Food Additives & Contaminants: Part A* 35, no. 1 (2018), 66–76. doi: https://doi.org/10.1080/19440049.2017.1397773.

[38] Li, K. Li, J.-J. Zhao, N. Ma, Y. and Di, B. "Removal of tetracycline in sewage and dairy products with high-stable MOF." *Molecules* 25, no. 6 (2020), 1312. doi: https://doi.org/10.3390/molecules25061312.

[39] Khatri, Jayraj, P. V. Nidheesh, T. S. Anantha Singh, and Suresh Kumar, M. "Advanced oxidation processes based on zero-valent aluminium for treating textile waste water." *Chemical Engineering Journal* 348 (2018), 67–73. doi: https://doi.org/10.1016/j.cej.2018.04.074.

[40] Robinson, T. McMullan, G. Marchant, R. and Nigam, P. "Remediation of dyes in textile effluent: a critical review on current treatment technologies with a proposed

[41] alternative." *Bioresource Technology* 77, no. 3 (2001), 247–255. doi: https://doi.org/10.1016/S0960-8524(00)00080-8.

[41] Tchinsa, A. Hossain, M. F. Wang, T. and Zhou, Y. "Removal of organic pollutants from aqueous solution using metal organic frameworks (MOFs)-based adsorbents: a review." *Chemosphere* 284 (2021), 131393. doi: https://doi.org/10.1016/j.chemosphere.2021.131393.

[42] Meteku, B. E. Huang, J. Zeng, J. Subhan, F. Feng, F. Zhang, Y. Qiu, Z. Aslam, S. Li, G. and Yan, Z. "Magnetic metal–organic framework composites for environmental monitoring and remediation." *Coordination Chemistry Reviews* 413 (2020), 213261. doi: https://doi.org/10.1016/j.ccr.2020.213261.

[43] Molavi, H. Hakimian, A. Shojaei, A. and Raeiszadeh, M. "Selective dye adsorption by highly water stable metal-organic framework: Long term stability analysis in aqueous media." *Applied Surface Science* 445 (2018), 424–436. doi: https://doi.org/10.1016/j.apsusc.2018.03.189.

[44] Eltaweil, A. S. Elshishini, H. M. Ghatass, Z. F. and Elsubruiti, G. M. "Ultra-high adsorption capacity and selective removal of Congo red over aminated graphene oxide modified Mn-doped UiO-66 MOF." *Powder Technology* 379 (2021), 407–416. doi: https://doi.org/10.1016/j.powtec.2020.10.084.

[45] Nanthamathee, C. and Dechatiwongse, P. "Kinetic and thermodynamic studies of neutral dye removal from water using zirconium metal-organic framework analogues." *Materials Chemistry and Physics* 258 (2021), 123924. doi: https://doi.org/10.1016/j.matchemphys.2020.123924.

[46] Ding, M. Cai, X. and Jiang, H-L. "Improving MOF stability: approaches and applications." *Chemical Science* 10, no. 44 (2019), 10209–10230. doi: https://doi.org/10.1039/C9SC03916C.

[47] Zhu, H. Zhang, Q. and Zhu, S. "Preparation of raspberry-like ZIF-8/PS composite spheres via dispersion polymerization." *Dalton Transactions* 44, no. 38 (2015), 16752–16757. doi: https://doi.org/10.1039/C5DT02627J.

[48] Meng, Y. Shu, L. Liu, L. Wu, Y. Xie, L-H. Zhao, M-J. and Li, J-R. "A high-flux mixed matrix nanofiltration membrane with highly water-dispersible MOF crystallites as filler." *Journal of Membrane Science* 591 (2019), 117360. doi: https://doi.org/10.1016/j.memsci.2019.117360.

[49] Yanai, N. Sindoro, M. Yan, J. and Granick, S. "Electric field-induced assembly of monodisperse polyhedral metal–organic framework crystals." *Journal of the American Chemical Society* 135, no. 1 (2013), 34–37. doi: https://doi.org/10.1021/ja309361d.

[50] Wang, Zhongguo, Lian Song, Yaquan Wang, Xiong-Fei Zhang, Dandan Hao, Yi F. and Yao, J. "Lightweight UiO-66/cellulose aerogels constructed through self-crosslinking strategy for adsorption applications." *Chemical Engineering Journal* 371 (2019), 138–144. doi: https://doi.org/10.1016/j.cej.2019.04.022.

[51] Wang, Z. Huang, J. Mao, J. Guo, Chen, Z. and Lai, Y. "Metal–organic frameworks and their derivatives with graphene composites: preparation and applications in electrocatalysis and photocatalysis." *Journal of Materials Chemistry A* 8, no. 6 (2020), 2934–2961. doi: https://doi.org/10.1039/C9TA12776C.

[52] Zhu, L. Zong, L. Wu, X. Li, M. H. Wang, J. Y. and Li. C. "Shapeable fibrous aerogels of metal–organic-frameworks templated with nanocellulose for rapid and large-capacity adsorption." *ACS Nano* 12, no. 5 (2018), 4462–4468. doi: https://doi.org/10.1021/acsnano.8b00566.

[53] Chen, Y. Li, S. Pei, X. Zhou, J. Feng, X. Zhang, S. Cheng, Y. Li, H. Han, R. and Wang, B. "A solvent-free hot-pressing method for preparing metal–organic-framework coatings." *Angewandte Chemie International Edition* 55, no. 10 (2016), 3419–3423. doi: https://doi.org/10.1002/anie.201511063.
[54] de Hatten, X. Bell, N. Yufa, N. Christmann, G. and Nitschke, J. R. "A dynamic covalent, luminescent metallopolymer that undergoes sol-to-gel transition on temperature rise." *Journal of the American Chemical Society* 133, no. 9 (2011), 3158–3164. doi: https://doi.org/10.1021/ja110575s.
[55] Zhu, H. Yang, X. Cranston, E. D. and Zhu, S. "Flexible and porous nanocellulose aerogels with high loadings of metal–organic-framework particles for separations applications." *Advanced Materials* 28, no. 35 (2016), 7652–7657. doi: https://doi.org/10.1002/adma.20160135.
[56] Ma, H. Wang, S. Meng, F. Xu, X. and Huo, X. "A hydrazone-carboxyl ligand-linked cellulose nanocrystal aerogel with high elasticity and fast oil/water separation." *Cellulose* 24, no. 2 (2017), 797–809. doi: 10.1007/s10570-016-1132-6.
[57] Zhang, W. Shi, S. Zhu, W. Huang, L. Yang, C. Li, S. Liu, X. et al. "Agar aerogel containing small-sized zeolitic imidazolate framework loaded carbon nitride: a solar-triggered regenerable decontaminant for convenient and enhanced water purification." *ACS Sustainable Chemistry & Engineering* 5, no. 10 (2017), 9347–9354. doi: https://doi.org/10.1021/acssuschemeng.7b02376.
[58] Bo, S. Ren, V Lei, C. Xie, Y. Cai, Y. Wang, S. Gao, J. Ni, Q. and Yao, J. "Flexible and porous cellulose aerogels/zeolitic imidazolate framework (ZIF-8) hybrids for adsorption removal of Cr (IV) from water." *Journal of Solid-State Chemistry* 262 (2018), 135–141. doi: https://doi.org/10.1016/j.jssc.2018.02.022.
[59] Leong, J-Y. Lam, W-H. Ho, K-W. Voo, W-P. Fu-Xiang Lee, M. Lim, H-P. Lim, S-L. Tey, B-T. Poncelet, D. and Chan, E-S. "Advances in fabricating spherical alginate hydrogels with controlled particle designs by ionotropic gelation as encapsulation systems." *Particuology* 24 (2016), 44–60. doi: https://doi.org/10.1016/j.partic.2015.09.004.
[60] Zhuang, Y. Kong, Y. Wang, X. and Shi, B. "Novel one step preparation of a 3D alginate-based MOF hydrogel for water treatment." *New Journal of Chemistry* 43, no. 19 (2019), 7202–7208. doi: https://doi.org/10.1039/C8NJ06031B.
[61] Cao, N. Lyu, Q. Li, J. Wang, Y. Yang, B. Szunerits, S. and Boukherroub, R. "Facile synthesis of fluorinated polydopamine/chitosan/reduced graphene oxide composite aerogel for efficient oil/water separation." *Chemical Engineering Journal* 326 (2017), 17–28. doi: https://doi.org/10.1016/j.cej.2017.05.117.
[62] Mao, J. Tang, Y. Wang, Y. Huang, J. Dong, X. Chen, Z. and Lai, Y. "Particulate matter capturing via naturally dried ZIF-8/graphene aerogels under harsh conditions." *IScience* 16 (2019), 133–144. doi: https://doi.org/10.1016/j.isci.2019.05.024.
[63] Torad, N. L. Hu, M. Ishihara, S. Sukegawa, H. Belik, A. A. Imura, M. Ariga, K. Sakka, Y. and Yamauchi, Y. "Direct synthesis of MOF-derived nanoporous carbon with magnetic Co nanoparticles toward efficient water treatment." *Small* 10, no. 10 (2014), 2096–2107. doi: https://doi.org/10.1002/smll.201302910.
[64] Liu, X. Li, S. Wang, D. Ma, Y. Liu, X. and Ning, M. "Theoretical study on the structure and cation–anion interaction of triethylammonium chloroaluminate ionic liquid." *Computational and Theoretical Chemistry* 1073 (2015), 67–74. doi: https://doi.org/10.1016/j.comptc.2015.09.012.

[65] Bhadra, B. Nath, J. Lee, Cho, C-W. and Jhung, S. H. "Remarkably efficient adsorbent for the removal of bisphenol A from water: Bio-MOF-1-derived porous carbon." *Chemical Engineering Journal* 343 (2018), 225–234. doi: https://doi.org/10.1016/j.cej.2018.03.004.

[66] Gupta, R. K. Dunderdale, G. J. England, M. W. and Hozumi, A. "Oil/water separation techniques: a review of recent progresses and future directions." *Journal of Materials Chemistry A* 5, no. 31 (2017), 16025–16058. doi: https://doi.org/10.1039/C7TA02070H.

[67] Antwi-Baah, R. and Liu, H. "Recent hydrophobic metal-organic frameworks and their applications." *Materials* 11, no. 11 (2018), 2250. doi: https://doi.org/10.3390/ma11112250.

[68] Roy, S. Suresh, V. M. and Kumar Maji, T. "Self-cleaning MOF: realization of extreme water repellence in coordination driven self-assembled nanostructures." *Chemical Science* 7, no. 3 (2016), 2251–2256. doi: https://doi.org/10.1039/C5SC03676C.

[69] Zhang, M. Xin, X. Xiao, Z. Wang, R. Zhang, L. and Sun, D. "A multi-aromatic hydrocarbon unit induced hydrophobic metal–organic framework for efficient C 2/C 1 hydrocarbon and oil/water separation." *Journal of Materials Chemistry A* 5, no. 3 (2017), 1168–1175. doi: https://doi.org/10.1039/C6TA08368D.

[70] Li, G. Zhao, Z. Liu, J. and Jiang, G. "Effective heavy metal removal from aqueous systems by thiol functionalized magnetic mesoporous silica." *Journal of Hazardous Materials* 192, no. 1 (2011), 277–283. doi: https://doi.org/10.1016/j.jhazmat.2011.05.015.

[71] Siddiqui, S. I. and Chaudhry, S. A. "Iron oxide and its modified forms as an adsorbent for arsenic removal: a comprehensive recent advancement." *Process Safety and Environmental Protection* 111 (2017), 592–626. doi: https://doi.org/10.1016/j.psep.2017.08.009.

[72] Xiao, R. Idris Abdu, H. Wei, L. Wang, T. Huo, S. Chen, J. and Lu, X. "Fabrication of magnetic trimetallic metal–organic frameworks for the rapid removal of tetracycline from water." *Analyst* 145, no. 6 (2020), 2398–2404. doi: https://doi.org/10.1039/C9AN02481F.

[73] Huang, L. He, M. Chen, B. and Hu, B. "A mercapto functionalized magnetic Zr-MOF by solvent-assisted ligand exchange for Hg^{2+} removal from water." *Journal of Materials Chemistry A* 4, no. 14 (2016), 5159–5166. doi: https://doi.org/10.1039/C6TA00343E.

4 Biopolymers Supported Nanomaterials for Water Treatment

Pragati Chauhan,[1] Sapna Nehra,[2]
Rekha Sharma,[1] and Dinesh Kumar[3]
[1]Department of Chemistry, Banasthali Vidyapith, Rajasthan 304022, India
[2]Department of Chemistry, Nirwan University, Jaipur, Rajasthan 303301, India
[3]School of Chemical Sciences, Central University of Gujarat, Gandhinagar 382030, India
*Corresponding author:
Email: sharma20rekha@gmail.com

CONTENTS

4.1 Introduction 63
4.2 Nanomaterials for Water Remediation 66
    4.2.1 Nanomaterials Derived from Polysaccharides for Water Treatment 66
        4.2.1.1 Cellulose-Based Nanomaterials 66
        4.2.1.2 Chitin/Chitosan-Based Nanomaterials 68
        4.2.1.3 Starch-Based Nanomaterials 70
        4.2.1.4 Gum-Based Nanomaterials 72
        4.2.1.5 Pectin Based Nanomaterials 73
4.3 Future Outlook 73
4.4 Conclusion 74
References 75

4.1 INTRODUCTION

Among the world's principal global ecological issues, potentially clean and potable water distribution is a hugely challenging fundamental issue worldwide. Nowadays, the rapidly increasing pollution of natural assets is a current issue that must be handled for the protection of the earth and its occupants in the future. The contamination of water after the industrial revolution became a major threat to drinking water. Rapid industrialization and enhanced anthropogenic activities have caused a sharp decline in freshwater assets, which were already limited. Natural toxins, for example, pharmaceutical drugs, heavy metals, and dyes, have become ecological threats. Among aqueous pollutants, cationic and anionic dyes are dominant dyes that threaten potable water. Methylene blue (MB) is a well-known product of, for example, wood, paper, and leather manufacture. Methyl orange (MO) is used in the food, printing, and paper industries, and the like. Congo red dye is used commonly in cotton textiles, in the paper industry (cellulose industry), and so forth. Crystal violet (CV) is used in bio industries such as for fingerprints, in dying ink, and the like. Conventional methods like electro-precipitation, natural process, reverse osmosis, chemical catalytic degradation, liquid membrane separation, coagulation/flocculation, ultra-filtration, adsorption, oxidation, sedimentation, and progressed oxidation measures are currently available advances in waste water treatment. Traditional strategies in treating waste water and cleaning can't provide a sufficient amount of purification achieve at a low enough cost [1].

The significant development in "green science" is catalysis with new handling frameworks and new arrangements of grouped catalysts having many advantages as far as item selectivity, measured usage, and a decrease in energy with the use of more stable substances and elective reaction media/conditions. Chemically active compounds have been used in recent waste water treatment methods, such as ozonation, photocatalysis, perovskite adsorption, and activated carbon filtration [2]. Biopolymers act as biosorbents, have minimal cost, and are inexhaustible. They can be used as alternatives to the costly energy-escalated commercially enacted carbons. Overall, for industries producing poisonous and dangerous chemically loaded waste water, the advancement of approaches to the eco-friendly treatment of waste water is a basic necessity [3]. Using nanostructures with different features such as significant chemical reactivity, large surface area, lesser power utilization, and cost-adequacy, could genuinely allow multifunctional nanosystems that enable molecule retention and the eradication of contaminations [4].

In the long term, manufactured polymers obtained through natural and non-renewable sources like petroleum and natural gas are reduced. Therefore, specialists are investigating more promptly accessible and economic options, specifically renewable natural polymers. Sustainable and environmentally friendly assets require natural polymers which might come from animals, plants, microbial biomass, algae, and monomeric parts [5]. These natural polymers have several fundamental features such as easy accessibility, security, biodegradability, and biocompatibility. Therefore, they are commonly used as nanosorbents. Cost-effective biopolymers, such as polysaccharides, are distinct in structure, size, and sub-atomic chains, making them appealing contenders for balance, immobilization, and decreasing NPs [6]. In contrast to traditional waste water systems, the advantages of natural polysaccharides include easy recovery and photocatalytic effectiveness.

Water bodies are contaminated due to the unregulated release of dye toxins from different processes such as in the manufacture of plastic, textile, paper, rubber, cosmetics, pharmaceutical, and in the food industry, prompting serious ecological questions [7]. Over 10,000 dyes are economically accessible worldwide, with around 7×10^5 tonnes used annually. About 12% of dyes are wasted, while 20% of these harm the environment via industrial waste water during manufacturing and other handling processes [8]. Every year, the textile and dye industries consume around 3×10^5 tonnes of manufactured toxic color effluents. The unregulated discharge of these colored toxic dyes into natural water bodies without effective treatment causes severe environmental pollution as they pollute groundwater systems and surface water. Though, even at the low level of 1.0 mg/L, they may seriously harm public wellbeing and the ecosystem [9]. Even as trace amounts photosynthesis in aquatic plants can be inhibited due to the diminishing of daylight penetration. Substances like heavy metals, aromatics, chloride, and the like, make the dyes toxic and thus harm the environment [10]. Dye waste is being released into water bodies without degradation and effective pre-treatment, causing hypersensitive dermatitis, skin issues, and even severe carcinogenic effects in humans and aquatic environments. Therefore, these toxic color effluents must be eliminated from water bodies. The advancement of eco-friendly handling approaches ought to be assumed as a basic component for the industries producing poisonous and dangerous chemically loaded waste water [11].

The division of various dyes is based on their method of preparation, their uses in various fields, their structures, and their chromogens as cationic (basic) dyes, nonionic (disperse) dyes, anionic (directly, acid and reactive) dyes [12]. Due to thermal stability, non-biodegradability, and photodegradation of the dyes, the treatment of polluted water is not simple. Thus, advancement in these novel materials is vital for efficient and financial economic dye elimination.

Physical, chemical, and biological methods include chemical catalytic degradation, electro-dialysis, biological treatments, chemical oxidation, liquid membrane separation, adsorption, electrolysis, coagulation, photodegradation, oxidation, and liquid chromatography, and so forth, all of which are recent methods to treat dye contaminated water [13]. These methods differ in their ecological impact, cost, and effectiveness. The physical adsorption process has stood out from the other methods because of easy availability, lesser cost, simplicity, no secondary pollution, and a more comprehensive range of applications [14]. Also, the adsorption method does not need a high working temperature and may eliminate many toxins instantaneously [15]. Adsorbents that are eco-friendly, easy, remarkably effective, accessible in large amounts, and meet the demands and standards of waste water treatment need sufficient exploration.

In recent years, bio-based nanocomposites have been actively explored in research because of their properties such as eco-friendliness, easily available, and effective adsorption capacity to treat waste water [16]. Polysaccharides are eco-friendly natural biopolymers produced from living organisms and long-chain monosaccharides ($C_n(H_2O)_n$). By comparing conventional adsorbents with polymeric materials, nanocomposite fabricates via polymeric materials offer the widest distribution, facility, environment-friendly process, and simple processing. These materials can be shaped into different structures such as films sheets, membranes, beads, and the like [17].

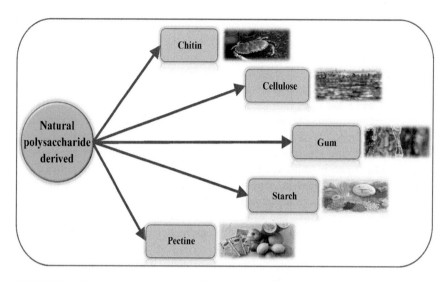

FIGURE 4.1 Renewable and eco-friendly natural polysaccharide(s).

Above all, they can desorb the dyes under altering conditions and can be regenerated. Functionalized natural polymeric materials, such as cellulose, chitin/chitosan, starch, pectin, and guar gum-based nanocomposites, have been very appropriate and eco-friendly adsorbents for the elimination of toxic dyes [18].

4.2 NANOMATERIALS FOR WATER REMEDIATION

4.2.1 NANOMATERIALS DERIVED FROM POLYSACCHARIDES FOR WATER TREATMENT

Nanomaterials derived from natural polysaccharides have shown excellent catalytic ability for waste water treatment. Figure 4.1 shows the different derivatives of renewable and eco-friendly natural polysaccharides. The present investigation pertains to the removal or adsorption of dyes through bio-based nanomaterials with the following objectives:

I. Cellulose-based nanomaterials
II. Chitin/chitosan-based nanomaterials
III. Starch-based nanomaterials
IV. Gum-based nanomaterials
V. Pectin-based nanomaterials

4.2.1.1 Cellulose-Based Nanomaterials

Natural polymer like cellulose has the widest distribution on earth in different forms such cellulose nanofibrils (CNFs) and cellulose nanocrystals (CNCs). Cellulose-based nanomaterials have shown various applications in pharmaceuticals, medicine, nanocomposites, and cosmetics. Because of the different properties of

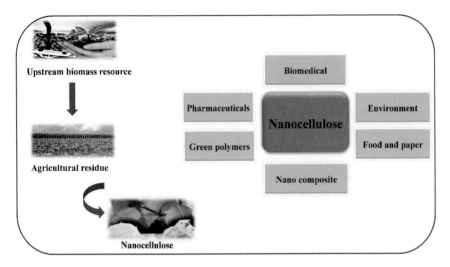

FIGURE 4.2 The significant role of cellulose in different fields.

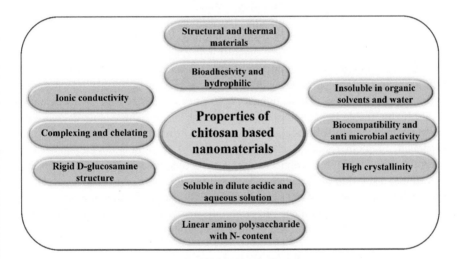

FIGURE 4.3 Properties of cellulose-based nanomaterials.

these cellulose-based nanomaterials, they have a significant role as adsorbents in other fields [19]. Figure 4.2 describes the various fields where we use nanocellulose.

Cellulose-based nanomaterials are suitable for waste water treatment because of their biodegradability and non-toxicity. A small volume of cellulose nanomaterials is required in medical applications. Therefore, their sustainability, cost, improved mechanical properties, life cycle, and suitable approach must be considered (Figure 4.3) [20]. Cellulose nanomaterials show ecological benefits over charcoal-derived activated carbon because it has less functionality.

CNFs and CNCs are used as catalysts or adsorbent materials when altered with anionic constituents to remove different cationic dyes. To increase their sorption capacity, carboxylation of cellulose-based nanomaterials can be used [21]. To form COO-modified or carboxylated CNCs, microcrystalline cellulose is hydrolyzed by ammonium persulfate (APS). At the time of cellulose hydrolysis, the carboxyl groups enter its surface. With the help of studies directed at the adsorption of cationic dyes, for example, MB affirmed the binding of positively charged dyes to the carboxylated group. The adsorption capacity of MB onto CNCs is 0.32 mmol/g after 10 min at 22°C. Desorption of MB using CNCs by ethanol can be over 90% but requires seven desorption cycles. But COO- modified CNCs, which can form in single-steps, can adsorb methyl blue by the entire UV degradation of MB in nearly 4 hrs. and has an increased efficiency rate compared to other CNCs. This shows that the addition of carboxyl groups that act as mandatory sites for the dyes causes modification of the surface of carboxylated CNCs. Different cellulose-based nanoadsorbents with contaminants, removal efficiencies, temperatures, and pHs are shown below in Table 4.1.

Covalent immobilization of CoPc was done onto nanofibers by BC to obtain CoPc@BC. CoPc@BC was used as a nanocatalyst and eliminated >90% of Rhodamine 6 B (RhB) dye from waste water within 3 hrs. using H_2O_2 as an oxidant. In another study, organic dyes like Rhodamine 6G and MB and MO are present in waste water, which gets eliminated using a polydopamine (PDA)/bmc hybrid membrane. This membrane catalyst gets reused ten times and easily gets isolated with no remarkable reduction in the efficiency of the catalyst [45]. CdS NPs get stabilized by coordination effects accompanied by bacterial CNFs through hydrothermal reactions. These hybrid photocatalyst CdS/BCNFs could degrade 82% MO by visible light irradiation exposure for 90 min. There is no decrease in the performance of the photocatalyst, and it could be used five times and easily get separated. In another study, nanocomposite $BC@TiO_2$, a new development [46] invented through 3D inter-connected porous BC blocks (namely, well-preserved) placed into a solution of Ti source, which has an expression of excellent potential, such as MO absorbent. Ag/TiO_2- and Au/TiO_2-cellulose CNF nanocomposites have been used for the adsorption and photocatalytic degradation of MB, and ~70% and 75% of MB, respectively, and can be removed within one hour.

4.2.1.2 Chitin/Chitosan-Based Nanomaterials

In nature, after cellulose, the second natural polymer that presents in ample amounts is chitosan/chitin. From the linear amino-polysaccharide of glucosamine, N-deacetylated chitin gets built up, which is the main component of crustaceous shells and insect cuticles. With the help of the chitin-deacetylase enzyme, chitin gets converted into chitosan through the enzymatic N-deacetylation reaction.

Unique physicochemical properties of chitosan-based nanomaterials are utilized to develop nanocatalysts and nanosorbents. The chitosan-based nanomaterials have been used to remove and adsorb dye molecules. The hydroxyl groups are viably used, and the amine groups are the most active group, namely, amine groups remain and impact other biobased polymer activities [47]. Dyes like acid green

TABLE 4.1
Various Cellulose-Based Biosorbents for Adsorption of Organic Dyes

Cellulose-Based Nanosorbents	Contaminant	Removal	Temperature (K)	pH	Reference
BC-AC	MB	505.8 mg/g	-	-	[22]
TiO_2/BC/PDA	MO, MB, RhB	95.1%, 99.5%, 100%	-	-	[23]
MoS_2-graphene-CFP	MB	485.4 mg/g	298	7.0	[24]
CA-PANI/β-CD nanofiber	MB	49.51 mg/g	298	8.0	[25]
IDA@CMC-PBQp microbeads	CV	107.52 mg/g	298	8.0	[26]
D-ZSMn/CNF, Cu- and Fe- ZSM/CNF	Rhodamine 6B, Reactive blue 4	34.36, 9.22, and 16.55 mg/g	-	7.0	[27]
CNF-Fe(0)@FeS	MB, CR	200.0, 111.1 mg/g	298	7.0, 5.0	[28]
CNCs/HPAM	MB	326.08 mg/g	298	5.0	[29]
CNCs/HPAM	MB	-	298	6.5	[30]
MnO_2 coated CNFs	MB	99.8%	298	9.6	[31]
PVAm microgels	CR 4BS	869.1 mg/g	298	3.5	[32]
CNC@polydopamine	MB	2066.72 mg/g	298	10.0	[33]
Cellulose nanocrystals	Victoria Blue 2B, MV 2B, rhodamine 6 G	98%, 90%, 78%	-	5.01	[34]
CMC/GOCOOH microbeads	MB	180.32 mg/g	298	10.0	[35]
CTAB modified CNC	CR	448.43 mg/g	298	7.5	[36]
CNC-ALG hydrogel bead	MB	255.5 mg/g	-	-	[37]
CNF-GnPrg aerogel	MB, CR	1178.5, 585.3 mg/g	-	-	[38]
CMC/g-C_3N_4/ZnO	MV	96.43 mg/g	298	8.0	[39]
G-C_3N_4/CNCs-H hydrogel	MB	232.558 mg/g	-	7.0	[40]
M_3D-PAA-CCNh	MB	332 mg/g	-	-	[41]
Carboxylated cellulose fabric filter	MB	76.92, 81.30 mg/g	298	5.0	[42]
PAETMAC-g-CNC	Neutral reactive blue 19 (RB 19)	>80%	298	7.0	[43]
Cellulose-modified $La_{0.9}Sr_{0.1}FeO_3$	CR	38.46 mg/g	-	4.0	[44]

25 adsorptions get favored by lessening the acetylation level of chitosan that boosts the related extent of amine groups available for protonation. The degree of acetylation and deacetylation is not proportional to the variation in these adsorption attributes. It differs because of dyes and the reconstruction of acetyl groups in the macromolecular chains, depending upon the preparation process [48].

CNF doped bio-hydrogels on CNCs show effective performance in eliminating contaminated water. Rear wire-like CNFs (60 to 120 nm) and semi-square CNCs (20 to 100 nm) were extracted from corn husks and shrimp shells. Various percentages of CNCs (0%, 1%, or 2%) of hybrid bio-aerogels (neat AR, AR1, and AR2) are synthesized, supported with CNFs through environmentally friendly freeze-drying methods.

Dyes such as RhB and MB removed by AR2 aerogel have been investigated. Due to their electrostatic interaction, they exhibit between CNCs (negatively charged) and dyes (positively charged) decorated AR2 accompanied by acetamide-enhanced groups help in the adsorption of toxic dyes. The hybrid bio-aerogels could be utilized five times with no detectable decrease in their catalytic performance. Magnetic chitosan nanocomposites that are cross-linked through glutaraldehyde showed considerable adsorption efficiency for removing dye such as Acid Red-2 with a maximum adsorption capacity of up to 91.6%. In contrast, iron oxide could adsorb only 16.4% [61]. Table 4.2 represents various chitin-based biosorbents that are used to eliminate various dyes.

In another study, along with the assistance of an external magnetic field, chitin/chitosan with a combination of Fe_3O_4 offered easy separation of toxic dyes. Carboxymethylated chitosan-conjugated nanosorbents comprising magnetic character showed huge adsorption capacity against Acid Green 25 and Crocein Orange G (1471 and 1883 mg/g, respectively) [62]. Different properties of chitosan-based nanomaterials are shown below in Figure 4.4.

4.2.1.3 Starch-Based Nanomaterials

Starch is another natural biopolymer, which comprises D-glucose units with bio-macromolecules comprising amylase, amylopectin, linear (1→4)-linked α-D-glucan, branched (1→6) α-D-glucan. It is extracted from different parts of plants: seeds, roots, and stalks. The main sources are rice, potatoes, maize, amongst others. Amongst other naturally occurring polymers, the nanocrystalline polymers have attracted a wide field of interest as nanosorbents due to their various attributes such as non-toxicity, biocompatibility, renewability, high surface area, low surface area cost, and high abundance. Compared to amorphous starch, nanocrystalline starch has been increasingly utilized to boost biopolymer [63].

Compared to native starch, modified starch shows increased water holding capabilities and reduced amylase solubility because of an elevation in crosslinking. For instance, epichlorohydrin is the most frequent crosslinking agent used for natural polysaccharides. In a related study, when corn starch becomes chemically bonded with epichlorohydrin, a bonded porous starch is produced. This then becomes hydrolyzed with α-amylase. This crosslinked porous starch was applied

TABLE 4.2
Chitin/Chitosan-Based Nanosorbents for Adsorption of Dyes

Chitin Based Nanomaterials	Contaminants	Temperature (°C)	pH	Reference
Urea modified TiO_2 doped Chitosan films	MG	300	-	[49]
Mesoporous silica/Nano-γ alumina/ Chitosan	MB	-	7.5	[50]
Nano-ZnO/chitosan microspheres	MO	37	-	[51]
PVF/CS/dopamine	Orange G	163.9	5.8 - 6.0	[52]
Fe_3O_4- chitosan NPs	Bromothymol Blue	24.85 - 44.85	3.0 - 7.4	[53]
$MnFe_2O_4$ impregnated chitosan-microspheres	MB	20 - 80	~ 6.0	[54]
3,5-DA /chitosan/$MnFe_2O_4$	MB	50 - 85	9.0	[55]
Chitosan ZnO_2 nano-beads	safranin	-	-	[56]
TiO_2 doped chitosan microspheres decorated onto cellulose acetate	MO	-	4.0	[57]
TiO_2 nanocomposite/ Chitosan tripolyphosphate	Reactive orange 16	40	4.0	[58]
chitosan-glutaraldehyde TiO_2 NPs	Reactive red 120	-	3.0 - 12	[59]
Chitosan ethylene glycol diglycidyl ether/TiO_2 NPs	Reactive orange 16	30 - 50	4.0 - 10	[60]

to adsorb MB. It shows an excellent adsorption capacity of 9.46 mg/g with an optimum temperature of 293K [64]. Moreover, a novel bioadsorbent, namely, TiO_2 NPs/CCMS, was synthesized using the sol-gel method. And then, TiO_2 NPs (~10 nm) were loaded onto the surface of crosslinked carboxymethyl starch (CCMS), showing photodegradation. These crosslinked nanosorbents were applied to adsorb cationic golden yellow/cationic yellow 28 X-GL dyes. Additionally, the (photo) nanocatalyst (TiO_2 NPs/CCMS), could be reused four times in successive cycles, without any observable loss in its efficiency [65].

Various cationic dyes can significantly adsorb by negatively charged starch. In another study, 3-Chloro-2-hydroxypropyl trimethyl-ammonium chloride, cornstarch, and epichlorohydrin (a cross-linked agent) were used to synthesize cross-linked cationic starch. In the adsorption of golden yellow SNE dye, the catalyst was also applied, and the resulting nanosorbents showed an excellent adsorption capacity of 208.77 mg/g with the optimum temperature at 308.15K [66].

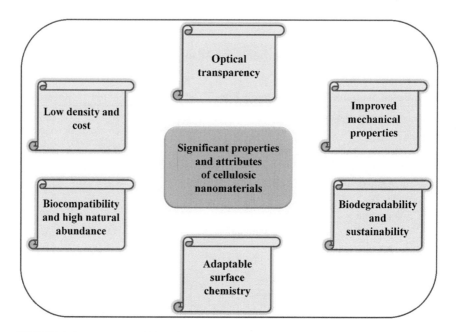

FIGURE 4.4 Properties of chitosan-based nanomaterials.

Nanocomposites MNPs@Starch-g-poly(vinyl sulfate) were synthesized using functionalized magnetic cross-linked starch materials and PVA. They were modified by vinyl acetate based on chlorosulfonic acid via copolymerization onto crude starch. The resultant biosorbents could successfully eliminate cationic dyes such as MG and MB. This sorbent shows adsorption capacities of 567 and 621 mg g^{-1}, respectively. Also, they can be reused for five successive cycles of adsorption-desorption [67].

4.2.1.4 Gum-Based Nanomaterials

Guar gum (GG) and gum arabic (GA) are green biopolymers and eco-friendly natural polysaccharides. Gum Arabic (GA), a branched heteropolysaccharide comprising D-galactopyranosyl units, is obtained from *Acacia senegal* trees. A mucoadhesive, viscous, water-soluble natural polysaccharide, namely, GG, was extracted from linear chains of D-galactopyranosyl units of the endosperm of guar beans. GA and GG have excellent potential in waste water treatment due to their significant properties such as ion exchange resins, thickening agents, and dispersing agents. They are effectively utilized to reduce or stabilize nanomaterials [68]. The magnetic NPs get effectively stabilized through these biosorbents and are used in various applications. Various nanocatalysts are deployed as emulsifying agents, potential stabilizers, and reducing agents in preparations.

Using guar gum-alginate blend (GGAlg@Ag) through the green route method AgNPs are synthesized, which are used to eliminate MB. The composition and chemical functionalities of GGAlg@Ag were confirmed by AFM, XRD, TEM, SEM, FTIR, and UV-vis techniques. A fusion related to CCD along with RSM analyzed optimized values. Also, it computed the system for different parameters such as pH (4.98), catalyst dose (0.07), exposure time (120 min), and dye concentration (194 mg/L), along with photodegradation capacity of 92.33% and desirability 1.0.

3D hydrogels are a source of an effective adsorbent material for water treatment because of their tremendous adsorption capacity and great (photo) catalytic degradation. Native GG by employing a crosslinking medium and self-assembly is used to prepare 3D biopolymer-based hydrogels. Properties like biodegradability, high surface area, biocompatibility, and high active-site availabilities show the ability in various applications of GG-based hydrogel nanocomposites [69]. For the photocatalytic degradation and adsorption of dye pollutants, a GG hybrid hydrogel-based nanocomposite, namely, NPs@MIL-100 (Fe)GG, was prepared through self-crosslinking and a simple blending process of MIL-100(Fe) with AgNPs.

4.2.1.5 Pectin Based Nanomaterials

Pectin is a linear polysaccharide that is obtained from the cell walls of higher plants with partially esterified polygalacturonic acid (PGA). Due to the non-toxic, biocompatible, and high molecular weight of these naturally occurring pectin-based nanosorbents, they are utilized in various applications. The structure of pectin during isolation varies with its origin and state, which is familiar to other polysaccharides [70]. Natural polymer pectin comprises D-galacturonic acid units and is combined in chains with α-(1-4) glycosidic linkage. The carboxyl hydroxyl groups are present along the backbone adjacent to neutral sugars as side chains. Carboxyl groups occur as methyl esters, but some are carboxamide groups when ammonia reacts. The resulting functional groups can fabricate compounds when combined with metal ions and do not require any toxic stabilizing agent, thus forming metal NPs. For example, the removal of methylene blue has been investigated through pectin–iron oxide magnetic adsorbent [71].

Several hybrid inorganic/organic nanocomposites are prepared via the combined incorporation of different fields. Various pectin-based nanocomposites used for adsorption or elimination of dyes are Fe-Pectin, Fe_3O_4 pectin, and FeNPs/Fe_3O_4NPs-CPA (crosslinked pectin adipic acid). Pectin-CuS nanocomposites (~50 nm) were synthesized through a facile co-precipitation and used in MB photodegradation at 30°C for 10 hrs. These nanosorbents can be effectively recycled ten times without any loss in their efficiency.

4.3 FUTURE OUTLOOK

With the expedient advancement in the investigation of bio-based NPs for dye adsorption in the last few years, we are as yet in the early days in this field, and some exploration with logical or useful significance is needed. More studies are required to upgrade the synthesis of natural/biopolymeric-based nanocomposites that are effective for industrial applications. To safeguard of every living being across the world, waste water treatment is the current need. There are potential uses of emerging nanotechnologies and their benefits in industrial waste water treatment via greener technology are enormous but need sufficient exploration. Through conventional approaches, the synthesis of polysaccharide-based nanomaterials could be easily accessible and safer for the ecosystem in terms of the conservation of resources. Therefore, they appear effective, eco-friendly, and sustainable alternatives for current treatment supplies.

Even though extensive studies in the catalytic and synthetic utilization of bio-based nanocomposites for the elimination of dye waste water have been done, there is still a particular consideration that requires exploration in future investigations to extend their utilization in various fields:

- More experiments should be performed on polysaccharide-based magnetic nanomaterials for dye adsorption.
- The fabrication of bionanomaterials and their application in waste water treatment requires different plant and animal remains, for example, husks, shells, or bones.
- Different parameters should be checked to fabricate biopolymer-based nanosorbents and their application in treating contaminated water bodies.
- A greener biological process needs to be promoted to improve the catalytic efficiency of these natural polymer-based nanomaterials.
- The appropriate range of biosorbents with relatively lower prices and excellent adsorption capability needs to be uplifted.

4.4 CONCLUSION

This chapter has explored bio-based nanocomposites for the adsorption of dyes. Natural polysaccharide-based nanocomposites are considered renewable, non-hazardous, eco-friendly, and effective for waste water treatment where it has been polluted by various toxic dyes. Biopolymers such as starch, chitin, gum, starch, and pectin supported Au, Pd, Cu, ZnO, Ag, TiO_2; Fe_3O_4 NPs are fabricated via green and straightforward methods. These biopolymers are supported because of their biological, biochemical, structural, and physiological features. They have properties such as biodegradability, non-toxicity, biocompatibility, economic viability, and abundant availability. Among the other natural polysaccharides, chitosan and cellulose are alternative renewable resources to conventional activated carbon because of their chemical stability, sustainability, abundance, and excellent physicochemical attributes towards toxic dye effluents. Surface reactivity, specific surface area, and maximum uptake capacity are different experimental parameters required for the effectiveness of biosorbents in terms of their adsorption capacities for specific metal species. Overall, natural biopolymer-based nanomaterials have been unquestionably the future direction for better utilization in waste water treatment, especially pollution by dyes that increasingly needs to be explored.

REFERENCES

[1] Khoramzadeh, E. Nasernejad, and Halladj, R. "Mercury biosorption from aqueous solutions by sugarcane bagasse." *Journal of the Taiwan Institute of Chemical Engineers* 44, no. 2 (2013): 266–269. https://doi.org/10.1016/j.jtice.2012.09.004

[2] Cai, Z. Dwivedi, A. D. Lee, W-N. Zhao, X. Liu, W. Sillanpää, M. D. Zhao, M. Huang, C-H. and Fu, J. "Application of nanotechnologies for removing pharmaceutically active compounds from water: development and future trends." *Environmental Science: Nano* 5, no. 1 (2018): 27–47. doi: 10.1039/C7EN00644F

[3] Crini, G. "Non-conventional low-cost adsorbents for dye removal: a review." *Bioresource Technology* 97, no. 9 (2006): 1061–1085. https://doi.org/10.1016/j.biortech.2005.05.001

[4] Meng, Da, Z. Zhu, L. Choi, J-G. Park, C-Y. and Oh, W-C. "Preparation, characterization and photocatalytic behavior of WO 3-fullerene/TiO2 catalysts under visible light." *Nanoscale Research Letters* 6, no. 1 (2011): 1–11. https://doi.org/10.1186/1556-276X-6-459

[5] Iravani, S. and Varma, R. S. "Greener synthesis of lignin nanoparticles and their applications." *Green Chemistry* 22, no. 3 (2020): 612–636. https://doi.org/10.1039/C9GC02835H

[6] Xu, C. Nasrollahzadeh, M. Selva, M. Issaabadi, Z. and Luque, R. "Waste-to-wealth: biowaste valorization into valuable bio (nano) materials." *Chemical Society Reviews* 48, no. 18 (2019): 4791–4822. https://doi.org/10.1039/C8CS00543E

[7] Alqadami, Abdullah, A. Naushad, M. Abdalla, M. A. Khan, M. R. and Alothman, Z. A. "Adsorptive removal of toxic dye using Fe3O4–TSC nanocomposite: equilibrium, kinetic, and thermodynamic studies." *Journal of Chemical & Engineering Data* 61, no. 11 (2016): 3806–3813. https://doi.org/10.1021/acs.jced.6b00446

[8] Essawy, A. A. A. Ali, El-Hag and Abdel-Mottaleb, M. S. A. "Application of novel copolymer-TiO2 membranes for some textile dyes adsorptive removal from aqueous solution and photocatalytic decolorization." *Journal of Hazardous Materials* 157, no. 2–3 (2008): 547–552. https://doi.org/10.1016/j.jhazmat.2008.01.072

[9] Xia, X. Zhou, Z. Wu, S. Wang, D. Zheng, S. and Wang, G. "Adsorption removal of multiple dyes using biogenic selenium nanoparticles from an Escherichia coli strain overexpressed selenite reductase CsrF." *Nanomaterials* 8, no. 4 (2018): 234. https://doi.org/10.3390/nano8040234

[10] Bharagava, Naresh, R. and Mishra, S. "Hexavalent chromium reduction potential of Cellulosimicrobium sp. isolated from common effluent treatment plant of tannery industries." *Ecotoxicology and Environmental Safety* 147 (2018): 102–109. https://doi.org/10.1016/j.ecoenv.2017.08.040

[11] Sivan, S. Kothaplamoottil, A. Padinjareveetil, K. K. Padil, V. V. T. Pilankatta, R. George, B. Senan, C. Černík, M. and Varma, R. S. "Greener assembling of MoO3 nanoparticles supported on gum arabic: cytotoxic effects and catalytic efficacy towards reduction of p-nitrophenol." *Clean Technologies and Environmental Policy* 21, no. 8 (2019): 1549–1561. https://doi.org/10.1007/s10098-019-01726-9

[12] Devi, Bala, N. and Mishra, S. "Solvent extraction equilibrium study of manganese (II) with Cyanex 302 in kerosene." *Hydrometallurgy* 103, no. 1–4 (2010): 118–123. https://doi.org/10.1016/j.hydromet.2010.03.007

[13] Pandey, S. Do, J. Y. Kim, J. and Kang, M. "Fast and highly efficient catalytic degradation of dyes using κ-carrageenan stabilized silver nanoparticles

[14] Pandey, S. "A comprehensive review on recent developments in bentonite-based materials used as adsorbents for waste water treatment." *Journal of Molecular Liquids* 241 (2017): 1091–1113. https://doi.org/10.1016/j.molliq.2017.06.115

[15] Zhang, J. Azam, M. S. Shi, C. Huang, J. Yan, B. Liu, Q. and Zeng, H. "Poly (acrylic acid) functionalized magnetic graphene oxide nanocomposite for removal of methylene blue." *RSC Advances* 5, no. 41 (2015): 32272–32282. https://doi.org/10.1039/C5RA01815C

[16] Thakur, S. Pandey, S. and Arotiba, O. A. "Development of a sodium alginate-based organic/inorganic superabsorbent composite hydrogel for adsorption of methylene blue." *Carbohydrate Polymers* 153 (2016): 34–46. https://doi.org/10.1016/j.carbpol.2016.06.104

[17] Crini, G. and Badot, P-M. "Application of chitosan, a natural aminopolysaccharide, for dye removal from aqueous solutions by adsorption processes using batch studies: A review of recent literature." *Progress in Polymer Science* 33, no. 4 (2008): 399–447. https://doi.org/10.1016/j.progpolymsci.2007.11.001

[18] Yusof, Y. M. and Kadir, M. F. Z. "Electrochemical characterizations and the effect of glycerol in biopolymer electrolytes based on methylcellulose-potato starch blend." *Molecular Crystals and Liquid Crystals* 627, no. 1 (2016): 220–233. https://doi.org/10.1080/15421406.2015.1137115

[19] Shatkin, J. A. Wegner, T. H. and Neih, W. "Incorporating life-cycle thinking into risk assessment for nanoscale materials: Case study of nanocellulose." *Production and Applications of Cellulose Nanomaterials* (2013): 89–112. doi: 10.1097/NT.0b013e3182435c79

[20] He, X. Male, K. B. Nesterenko, P. N. Brabazon, D. Paull, B. and Luong, J. H. T. "Adsorption and desorption of methylene blue on porous carbon monoliths and nanocrystalline cellulose." *ACS Applied Materials & Interfaces* 5, no. 17 (2013): 8796–8804. https://doi.org/10.1021/am403222u

[21] Khamkeaw, A. Jongsomjit, B. Robison, J. and Phisalaphong, M. "Activated carbon from bacterial cellulose as an effective adsorbent for removing dye from aqueous solution." *Separation Science and Technology* 54, no. 14 (2019): 2180–2193. https://doi.org/10.1080/01496395.2018.1541906

[22] Yang, L. Chen, C. Hu, Y. Wei, F. Cui, J. Zhao, Y. Xu, X. Chen, X. and Sun, D. "Three-dimensional bacterial cellulose/polydopamine/TiO2 nanocomposite membrane with enhanced adsorption and photocatalytic degradation for dyes under ultraviolet-visible irradiation." *Journal of Colloid and Interface Science* 562 (2020): 21–28. https://doi.org/10.1016/j.jcis.2019.12.013

[23] Gopalakrishnan, A. Singh, S. P. and Badhulika, S. "Reusable, few-layered-MoS 2 nanosheets/graphene hybrid on cellulose paper for superior adsorption of methylene blue dye." *New Journal of Chemistry* 44, no. 14 (2020): 5489–5500. https://doi.org/10.1039/D0NJ00246A

[24] Ali, A. El-Aassar, S. M. M. R. Hashem, F. S. and Moussa, N. A. "Surface modified of cellulose acetate electrospun nanofibers by polyaniline/β-cyclodextrin composite for removal of cationic dye from aqueous medium." *Fibers and Polymers* 20, no. 10 (2019): 2057–2069. https://doi.org/10.1007/s12221-019-9162-y

[25] Omer, A. M. Elgarhy, G. S. El-Subruiti, G. M. Khalifa, R. E. and Eltaweil, A. S. "Fabrication of novel iminodiacetic acid-functionalized carboxymethyl

cellulose microbeads for efficient removal of cationic crystal violet dye from aqueous solutions." *International Journal of Biological Macromolecules* 148 (2020): 1072–1083. https://doi.org/10.1016/j.ijbiomac.2020.01.182

[26] Lakhane, M. Mahabole, M. Bogle, K. Khairnar, R. and Kokol, V. "Nanocomposite films prepared from differently modified ZSM-5 zeolite and cellulose nanofibrils for cationic and anionic dyes removal." *Fibers and Polymers* 20, no. 10 (2019): 2127–2139. https://doi.org/10.1007/s12221-019-1139-3

[27] Qian, X. Xu, Y. Yue, X. Wang, C. Liu, M. Duan, C. Xu, Y. Zhu, C. and Dai, L. "Microwave-assisted solvothermal in-situ synthesis of CdS nanoparticles on bacterial cellulose matrix for photocatalytic application." *Cellulose* 27 (2020): 5939–5954. DOI: 10.4236/ajac.2018.94017

[28] Zhou, C. Wu, Q. Lei, T. and Negulescu, I. I. "Adsorption kinetic and equilibrium studies for methylene blue dye by partially hydrolyzed polyacrylamide/cellulose nanocrystal nanocomposite hydrogels." *Chemical Engineering Journal* 251 (2014): 17–24. https://doi.org/10.1016/j.cej.2014.04.034

[29] Zhou, C. Lee, S. Dooley, K. and Wu, Q. "A facile approach to fabricate porous nanocomposite gels based on partially hydrolyzed polyacrylamide and cellulose nanocrystals for adsorbing methylene blue at low concentrations." *Journal of Hazardous Materials* 263 (2013): 334–341. https://doi.org/10.1016/j.jhazmat.2013.07.047

[30] Wang, Y. Yadav, S. Heinlein, T. Konjik, V. Breitzke, H. Buntkowsky, G. Schneider, J. J. and Zhang, K. "Ultra-light nanocomposite aerogels of bacterial cellulose and reduced graphene oxide for specific absorption and separation of organic liquids." *RSC Advances* 4, no. 41 (2014): 21553–21558. doi: 10.1039/C4RA02168A

[31] Jin, L. Sun, Q. Xu, Q. and Xu, Y. "Adsorptive removal of anionic dyes from aqueous solutions using microgel based on nanocellulose and polyvinylamine." *Bioresource Technology* 197 (2015): 348–355. https://doi.org/10.1016/j.biortech.2015.08.093

[32] Wang, G. Zhang, J. Lin, S. Xiao, H. Yang, Q. Chen, S. Yan, B. and Gu, Y. "Environmentally friendly nanocomposites based on cellulose nanocrystals and polydopamine for rapid removal of organic dyes in aqueous solution." *Cellulose* 27, no. 4 (2020): 2085–2097. https://doi.org/10.1007/s10570-019-02944-6

[33] Karim, Z. Mathew, A. P. Grahn, M. Mouzon, J. and Oksman, K. "Nanoporous membranes with cellulose nanocrystals as functional entity in chitosan: removal of dyes from water." *Carbohydrate Polymers* 112 (2014): 668–676. https://doi.org/10.1016/j.carbpol.2014.06.048

[34] Eltaweil, A. S. Elgarhy, G. S. El-Subruiti, G. M. and Omer, A. M. "Carboxymethyl cellulose/carboxylated graphene oxide composite microbeads for efficient adsorption of cationic methylene blue dye." *International Journal of Biological Macromolecules* 154 (2020): 307–318. https://doi.org/10.1016/j.ijbiomac.2020.03.122

[35] Ranjbar, D. Raeiszadeh, M. Lewis, L. MacLachlan, M. J. and Hatzikiriakos, S. G. "Adsorptive removal of Congo red by surfactant modified cellulose nanocrystals: a kinetic, equilibrium, and mechanistic investigation." *Cellulose* 27, no. 6 (2020): 3211–3232. https://doi.org/10.1007/s10570-020-03021-z

[36] Mohammed, N. Grishkewich, N. Waeijen, H. A. Berry, R. M. and Tam, K. C. "Continuous flow adsorption of methylene blue by cellulose nanocrystal-alginate

[37] Yu, Z. Hu, C. Dichiara, A. B. Jiang, W. and Gu, J. "Cellulose nanofibril/carbon nanomaterial hybrid aerogels for adsorption removal of cationic and anionic organic dyes." *Nanomaterials* 10, no. 1 (2020): 169. https://doi.org/10.3390/nano10010169

[38] Sharma, G. Kumar, A. Sharma, S. Naushad, M. Ghfar, A. A. Ala'a, H. Ahamad, T. Sharma, N. and Stadler, F. J. "Carboxymethyl cellulose structured nano-adsorbent for removal of methyl violet from aqueous solution: Isotherm and kinetic analyses." *Cellulose* 27, no. 7 (2020): 3677–3691. https://doi.org/10.1007/s10570-020-02989-y

[39] Wang, H. Li, J. Ding, N. Zeng, X. Tang, X. Sun, Y. Lei, T. and Lin, L. "Eco-friendly polymer nanocomposite hydrogel enhanced by cellulose nanocrystal and graphitic-like carbon nitride nanosheet." *Chemical Engineering Journal* 386 (2020): 124021. https://doi.org/10.1016/j.cej.2020.124021

[40] Samadder, R. Akter, N. Roy, A. C. Uddin, M. M. Hossen, M. J. and Azam, M. S. "Magnetic nanocomposite based on polyacrylic acid and carboxylated cellulose nanocrystal for the removal of cationic dye." *RSC Advances* 10, no. 20 (2020): 11945–11956. doi: 10.1039/D0RA00604A

[41] Li, C. Ma, H. Venkateswaran, S. and Hsiao, B. S. "Highly efficient and sustainable carboxylated cellulose filters for removal of cationic dyes/heavy metals ions." *Chemical Engineering Journal* 389 (2020): 123458. https://doi.org/10.1016/j.cej.2019.123458

[42] Jiang, X. Lou, C. Hua, F. Deng, H. and Tian, X. "Cellulose nanocrystals-based flocculants for high-speed and high-efficiency decolorization of colored effluents." *Journal of Cleaner Production* 251 (2020): 119749. https://doi.org/10.1016/j.jclepro.2019.119749

[43] Ali, S. M. and Al-Oufi, B. "Synergistic sorption performance of cellulose-modified La 0.9 Sr 0.1 FeO3 for organic pollutants." *Cellulose* 27, no. 1 (2020): 429–440. https://doi.org/10.1007/s10570-019-02814-1

[44] Chen, S. and Huang, Y. "Bacterial cellulose nanofibers decorated with phthalocyanine: preparation, characterization and dye removal performance." *Materials Letters* 142 (2015): 235–237. https://doi.org/10.1016/j.matlet.2014.12.036

[45] Gholami D. H. Jiang, Q. Ghim, D. Cao, S. Chandar, Y. J. Morrissey, J. J. Jun, Y-S. and Singamaneni, S. "A robust and scalable polydopamine/bacterial nanocellulose hybrid membrane for efficient waste water treatment." *ACS Applied Nano Materials* 2, no. 2 (2019): 1092–1101. https://doi.org/10.1021/acsanm.9b00022

[46] Zhang, Z. Ma, X. Jia, M. Li, B. Rong, J. and Yang, X. "Deposition of CdTe quantum dots on microfluidic paper chips for rapid fluorescence detection of pesticide 2, 4-D." *Analyst* 144, no. 4 (2019): 1282–1291. doi: 10.1039/C8AN02051E

[47] Saha, T. K. Ichikawa, H. and Fukumori, Y. "Gadolinium diethylenetriaminopentaacetic acid-loaded chitosan microspheres for gadolinium neutron-capture therapy." *Carbohydrate Research* 341, no. 17 (2006): 2835–2841. https://doi.org/10.1016/j.carres.2006.09.016

[48] Rinaudo, M. "Chitin and chitosan: properties and applications." *Progress in Polymer Science* 31, no. 7 (2006): 603–632. https://doi.org/10.1016/j.progpolymsci.2006.06.001

[49] Pérez-Obando, J. Marin-Silva, D. A. Pinotti, A. N. Pizzio, L. R. Osorio-Vargas, P. and Rengifo -Herrera, J. A. "Degradation study of malachite green on chitosan films containing heterojunctions of melon/TiO2 absorbing visible-light in solid-gas interfaces." *Applied Catalysis B: Environmental* 244 (2019): 773–785. https://doi.org/10.1016/j.apcatb.2018.12.004

[50] Zhong, R. Zhong, Q. Huo, M. Yang, B. and Li, H. "Preparation of biocompatible nano-ZnO/chitosan microspheres with multi-functions of antibacterial, UV-shielding and dye photodegradation." *International Journal of Biological Macromolecules* 146 (2020): 939–945. https://doi.org/10.1016/j.ijbiomac.2019.09.217

[51] Zhang, G. Li, Y. Gao, A. Zhang, Q. Cui, J. Zhao, S. Zhan, X. and Yan, Y. "Bio-inspired underwater superoleophobic PVDF membranes for highly-efficient simultaneous removal of insoluble emulsified oils and soluble anionic dyes." *Chemical Engineering Journal* 369 (2019): 576–587. https://doi.org/10.1016/j.cej.2019.03.089

[52] Sahbaz, D. A. Yakar, A. and Gündüz, U. "Magnetic Fe3O4-chitosan micro-and nanoparticles for waste water treatment." *Particulate Science and Technology*, 37, 732-740, (2018). https://doi.org/10.1080/02726351.2018.1438544

[53] Jyothi, M. S. V. Angadi, J. Kanakalakshmi, T. V. Padaki, M. Geetha, B. R. and Soontarapa, K. "Magnetic nanoparticles impregnated, crosslinked, porous chitosan microspheres for efficient adsorption of methylene blue from pharmaceutical waste water." *Journal of Polymers and the Environment* 27, no. 11 (2019): 2408–2418. https://doi.org/10.1007/s10924-019-01531-x

[54] Shoueir, K. El-Sheshtawy, H. Misbah, M. El-Hosainy, H. El-Mehasseb, I. and El-Kemary, M. "Fenton-like nanocatalyst for photodegradation of methylene blue under visible light-activated by hybrid green DNSA@ Chitosan@ MnFe$_2$O$_4$." *Carbohydrate Polymers* 197 (2018): 17–28. https://doi.org/10.1016/j.carbpol.2018.05.076

[55] Roshitha, Sundaram, S. Mithra, V. Saravanan, V. Sadasivam, S. K. and Gnanadesigan, M. "Photocatalytic degradation of methylene blue and safranin dyes using chitosan zinc oxide nano-beads with Musa× paradisiaca L. pseudo stem." *Bioresource Technology Reports* 5 (2019): 339–342. https://doi.org/10.1016/j.biteb.2018.08.004

[56] Shi, X. Zhang, X. Ma, L. Xiang, C. and Li, L. "TiO$_2$-doped chitosan microspheres supported on cellulose acetate fibers for adsorption and photocatalytic degradation of methyl orange." *Polymers* 11, no. 8 (2019): 1293. https://doi.org/10.3390/polym11081293

[57] Abdulhameed, A. S. Mohammad, A. K-T. and Jawad, A. H. "Application of response surface methodology for enhanced synthesis of chitosan tripolyphosphate/TiO2 nanocomposite and adsorption of reactive orange 16 dye." *Journal of Cleaner Production* 232 (2019): 43–56. https://doi.org/10.1016/j.jclepro.2019.05.291

[58] Jawad, A. H. Mubarak, N. S. A. and Abdulhameed. A. S. "Tunable Schiff's base-cross-linked chitosan composite for the removal of reactive red 120 dye: adsorption and mechanism study." *International Journal of Biological Macromolecules* 142 (2020): 732–741. https://doi.org/10.1016/j.ijbiomac.2019.10.014

[59] Abdulhameed, A. S. Jawad, A. H. and Mohammad, A. K-T. "Synthesis of chitosan-ethylene glycol diglycidyl ether/TiO2 nanoparticles for adsorption of reactive

orange 16 dye using a response surface methodology approach." *Bioresource Technology* 293 (2019): 122071. https://doi.org/10.1016/j.biortech.2019.122071

[60] Kadam, A. A. and Lee, D. S. "Glutaraldehyde crosslinked magnetic chitosan nanocomposites: Reduction precipitation synthesis, characterization, and application for removal of hazardous textile dyes." *Bioresource Technology* 193 (2015): 563–567. https://doi.org/10.1016/j.biortech.2015.06.148

[61] Chang, Chuang, Y. and Chen, D-H. "Adsorption kinetics and thermodynamics of acid dyes on a carboxymethylated chitosan-conjugated magnetic nano-adsorbent." *Macromolecular Bioscience* 5, no. 3 (2005): 254–261. https://doi.org/10.1002/mabi.200400153

[62] Visakh, P. M. Mathew, A. J. I. P. Oksman, K. and Thomas, S. "Starch-based bionanocomposites: processing and properties." *Polysaccharide Building Blocks: A Sustainable Approach to the Development of Renewable Biomaterials* (2012): 287–306. doi: 10.1002/9781118229484.ch11

[63] Angellier, H. Molina-Boisseau, S. and Dufresne, A. "Mechanical properties of waxy maize starch nanocrystal reinforced natural rubber." *Macromolecules* 38, no. 22 (2005): 9161–9170. https://doi.org/10.1021/ma0512399

[64] Guo, L. Li, G. Liu, J. Meng, Y. and Tang, Y. "Adsorptive decolorization of methylene blue by crosslinked porous starch." *Carbohydrate Polymers* 93, no. 2 (2013): 374–379. https://doi.org/10.1016/j.carbpol.2012.12.019

[65] Huang, J. Chang, P. R. Lin, N. and Dufresne, A. "Polysaccharide-based nanocrystals: chemistry and applications." *John Wiley & Sons,* 8, No. 392, 1-33, (2014). https://doi.org/10.3389/fchem.2020.00392

[66] Guo, J. Wang, J. Zheng, G. and Jiang, X. "Optimization of the removal of reactive golden yellow SNE dye by crosslinked cationic starch and its adsorption properties." *Journal of Engineered Fibers and Fabrics* 14 (2019): 1558925019865260. https://doi.org/10.1177%2F1558925019865260

[67] Sharma, R. Kalia, S. Kaith, B. S. Pathania, D. Kumar, A. and Thakur, P. "Guaran-based biodegradable and conducting interpenetrating polymer network composite hydrogels for adsorptive removal of methylene blue dye." *Polymer Degradation and Stability* 122 (2015): 52–65. https://doi.org/10.1016/j.polymdegradstab.2015.10.015

[68] Duan, C. Liu, C. Meng, X. Gao, K. Lu, W. Zhang, Y. Dai, L. et al. "Facile synthesis of Ag NPs@ MIL-100 (Fe)/guar gum hybrid hydrogel as a versatile photocatalyst for waste water remediation: photocatalytic degradation, water/oil separation and bacterial inactivation." *Carbohydrate Polymers* 230 (2020): 115642. https://doi.org/10.1016/j.carbpol.2019.115642

[69] Ridley, B. L. O'Neill, M. A. and Mohnen, D. "Pectins: structure, biosynthesis, and oligogalacturonide-related signaling." *Phytochemistry* 57, no. 6 (2001): 929–967. https://doi.org/10.1016/S0031-9422(01)00113-3

[70] Rakhshaee, R. and Panahandeh, M. "Stabilization of a magnetic nano-adsorbent by extracted pectin to remove methylene blue from aqueous solution: a comparative studying between two kinds of cross-likened pectin." *Journal of Hazardous Materials* 189, no. 1–2 (2011): 158–166. https://doi.org/10.1016/j.jhazmat.2011.02.013

[71] Liu, Fu, J. Zhao, Z-S. and Jiang, G-B. "Coating Fe3O4 magnetic nanoparticles with humic acid for highly efficient removal of heavy metals in water." *Environmental Science & Technology* 42, no. 18 (2008): 6949–6954. https://doi.org/10.1021/es800924c

5 Disinfection of Water by Eco-friendly Nanomaterials

Poonam Ojha[1] and Meena Nemiwal[2]*
[1]Department of Chemistry, Swami Keshvanand Institute of Technology, Management and Gramothan, Jaipur, India 302017
[2]Department of Chemistry, Malaviya National Institute of Technology, Jaipur, India 302017
*Corresponding author: Email: meena.chy@mnit.ac.in

CONTENTS

5.1 Introduction 81
5.2 Nanomaterials for the Treatment of Water 83
5.3 Environmental-Friendly Nanomaterials 84
5.4 Plant-Based Materials for the Synthesis of Metal and Metal Oxide NPs 86
5.5 Microbes for the Synthesis of Metal Nps 87
5.6 Biopolymer-Based Nanomaterials 87
5.7 The Mechanism of Disinfection by Ecofriendly Nanomaterials ... 89
    5.7.1 Disinfection by Plant-Based NPs 89
5.8 Conclusion 92
References 92

5.1 INTRODUCTION

A most essential and unique resource in our ecosystem is water. Water is not only a universal solvent, but it is also a crucial resource necessary for living beings on earth. Though it is the most abundant natural source, for our consumption, only 1% of it is available [1]. Because of the growing population and an exponential rise in anthropogenic activities, freshwater resources are being depleted in quality because of contamination by various organic impurities. Examples include animal and plant waste, domestic waste irrigation, agricultural waste such as pesticides and inorganic chemical fertilizers, heavy metals such as cadmium, mercury, iron, lead, and non-metallic ions such as chloride and fluoride [2]. The exposure of the environment to these pollutants is partly due to the illegal management of resources. So, water pollution has multiplied many times and waterborne infections worldwide have intensified in the last few years [3].

The microbiological pollution and pathogens in water bodies are an issue of serious concern. Waterborne pathogens create worldwide challenges, and thus, millions of people worldwide lack safe and consumable water for survival. As per UNICEF, because of waterborne pathogens, diseases such as cholera, dysentery, hepatitis A, Salmonella giardia, and so forth, are caused, and millions of people lose their lives [4]. These waterborne pathogens significantly contaminate both surface water bodies and underground water resources, and thus they are a solemn menace for human health. So, the purification and disinfection of water have become one of the biggest challenges for the modern world [5]. Thus, major attention is required to restrict the biological contamination of water and search for efficient methods to reduce the growth of pathogens and the harmful effects caused by them.

The first municipal water treatment plant was established in 1804 in Scotland. In this method, the sand filter carried out plant filtration of impure water. After that, Humphrey Davy, in 1814, introduced the most basic mode of disinfection using chlorine, which was further accepted as an application in the water disinfection process around 1902. Soon after, a novel corridor for water disinfection was discovered. It was killing microorganisms using ozone. This method was known as "ozonation." Since then, many modern methods for water treatment, disinfection, and the treatment of waterborne pathogens have been discovered. Among all other methods, chlorination is the most accepted and widely used method as it is the most economical and straightforward method [6, 7].

The most common method applied for removing pathogens, especially bacteria and fungi, from the system is adsorption. In most adsorption techniques, viruses pass through filter beds as they are tiny. With time, some pathogens such as cryptosporidium giardia, and so forth, have become resistant to disinfectants. Thus, large quantities of these chemicals are required to be added to water to kill pathogens, increasing the concentration of by products in water systems. Economic equipment and efficient separation methods are needed to remove disease-causing microorganisms from water and a new class of eco-friendly adsorbents and disinfectants. But unfortunately, chemical disinfectants cannot serve the whole purpose of complete sterilization and disinfection as they are non-biodegradable and leave toxic contaminants in water. Above all, conventional water purification and disinfection methods face many other challenges such as costly equipment, slow processing, problems faced during separation, and the final disposal of filtration waste, limits their applications. Thus, new environmentally friendly, efficient and power technologies are required for municipal and waste-water treatments to fulfill the ever-increasing demand for safe and good quality water supply. Although many water purification technologies have been discovered in past years, the evolution of nanomaterials for water purification is an unusual development [8–11]. The present chapter discusses the disinfection of water by nanomaterials synthesized by the green route. Synthesis of nanomaterials can be

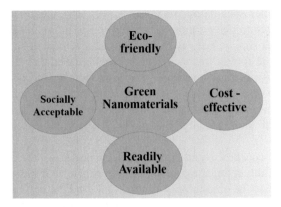

FIGURE 5.1 Advantages of green nanomaterials.

carried out with the help of materials such as plant extract, microbes, and the like, which may replace the use of harmful chemical reagents. In addition, plant-based nanomaterials are a sustainable way for the disinfection of water. Figure 5.1 shows the benefits of green nanomaterials.

5.2 NANOMATERIALS FOR THE TREATMENT OF WATER

Nanomaterials have emerged as a new field in engineering, science, and technology. It is the science of the manner of handling materials on a nanoscale. Nanomaterials are materials whose components are sized at dimensions between 1 and 100 nm. They are tiny, in other words, nano. Their physical, mechanical, electrical, and optical activities differ significantly from conventional bigger materials [12]. They are highly reactive and show characteristic features such as catalysis and adsorption. They are found extensively and with distinct properties. In recent years, nanomaterials have found their applications in purification, disinfection, and water remediation for municipal purposes. Nanomaterials have provided an abundance of solutions in water purification and remediation technologies. Also, new emerging materials such as metal-organic frameworks and covalent organic frameworks have also been found to play a crucial role in water and environmental remediation [13 - 18]. Nanomaterials used in water treatment methods are very effective in removing disease-causing microorganisms and various pathogens. Thus, based on various studies, nanomaterials used in water and waste-water treatment are a novel substitute for conventional water treatment methods. Modern advances in nanotechnology methods propose remarkable opportunities to create a new generation of water supply and purification systems.

The use of NPs in water purification methods seems very promising. Nanomaterials can be used in various stages of water purification such as microfiltration, ultrafiltration, nanofiltration, or reverse osmosis to remove various kinds of sediments, chemical contaminants, ions, trace elements, and other microbial impurities such as microorganisms [19]. This clearly shows that a lot of effort is required to deliver more successful routes for maintaining the highest standards of water supply, keeping in mind the achievement of scalability and the effectiveness of the applied methods. It must be understood that the connotation of the nanotechnological tools developed so far cannot be underestimated since they are of great value when it comes to pioneering water treatment methods and raising awareness of the basic mechanisms of such processes [20]. Current progress in nanotechnological tools predicts extraordinary opportunities to expand water supply systems.

There is no concern regarding the effectiveness of the deployment of nanomaterials in the disinfection of water. Nevertheless, it has some important issues that need to be minimized, as these NPs might get released into the environment and accumulate. This toxic release can cause severe effects on human health and can degrade the environment [21]. So, while using nanomaterials, their toxicity should be evaluated to reduce health risks. Also, while modifying existing nanomaterials, their ecotoxicity must be revaluated. Thus, an overall assessment of the effects of nanomaterials on the environment at every stage, namely, a life cycle assessment, should be carried out to calculate overall profits and threats.

Further research is needed to develop environmental-friendly and economical methods for synthesizing nanomaterials. Their efficiency in water treatment technology on a large scale is a matter of great concern. The mechanism of antimicrobial resistance by NPs includes interaction with genetics at the DNA level and biochemistry at the protein level of bacteria. Nanoparticles interrupt the formation of the bacterial biofilms and can enter through the cell wall and membrane of bacteria and disturbs the molecular mechanisms. [22].

5.3 ENVIRONMENTAL-FRIENDLY NANOMATERIALS

As we discuss beyond, there are various areas of concern in the application of nanotechnology in the field. Such issues necessitate careful examination and should be appropriately analyzed. A focus on preparing environmentally sustainable nanomaterials is needed in order to help with the issue of water scarcity. Economical methods should be researched to overcome the possible toxicity and other issues related to nanomaterials. A new category of nanomaterials has been found along these lines, namely, eco-friendly nanomaterials. In this context, green chemistry principles have been developed for the synthesis of nanomaterials. The following principles need to be followed: (1) the use of toxic reagents, and organic solvents are avoided, (2) there should be no by-product formation during synthesis, (3) particle shape and size and dispersion are controlled, (4) a purified final product is formed. Natural bio-based nanomaterials are synthesized through various biomolecules such as carbohydrates (chitosan, glycogens, and polysaccharides), plants, and bacterial cells. Biodegradable nanomaterials such as biopolymers, bionanoparticles synthesized from plant extracts, nanocellulose, and the like, can

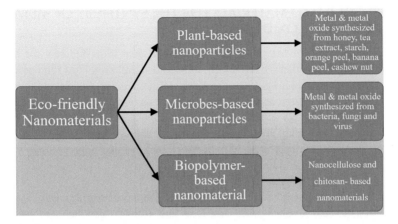

FIGURE 5.2 Classification and types of eco-friendly nanomaterials.

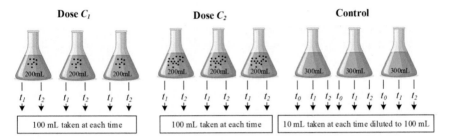

FIGURE 5.3 Schematic for measuring antibacterial procedure for various microorganisms, nanoparticle type, and water type (Adapted with permission from [24]. Copyright (2019) Copyright the Authors, some rights reserved; exclusive licensee [MDPI]. Distributed under a Creative Commons Attribution License 4.0 (CC BY)).

play an essential role in saving our environment by evading secondary pollutants. Figure 5.2 shows the classification and types of eco-friendly nanomaterials. Such nanomaterials have fewer effects on the environment and human lives and can be used significantly in everyday life. A proper scrutiny while using them in laboratories and different techniques should also be accurately carried out [23].

In this regard, Al-Issai et al. developed different NPs of silver, copper, silver-copper, zinc oxide, magnesium oxide, silicon dioxide, and carbon nanotubes. They investigated their antimicrobial activity by batch studies in desalinated water samples containing E. coli, Enterobacter, Salmonella, and Enterococci. Enhanced disinfection activity was observed for all the bacteria Figure 5.3 [24]. A brief study of various types of nanomaterials synthesized by environmental-friendly methods shows their potential applications in the disinfection of water:

5.4 PLANT-BASED MATERIALS FOR THE SYNTHESIS OF METAL AND METAL OXIDE NPS

Greener methods involving plant-based materials seem to be the best for large-scale single pot biosynthesis of NPs [25]. Plant parts (leaf, root, latex) and products derived from them, such as extracts from tea, coffee, banana, simple amino acids, wine, table sugar, and glucose, have been utilized as reagents. A significant effort is made with tea extracts as they have a high amount of polyphenols which can act as chelating, reducing, and capping agents for the synthesis of NPs [26]. The resulting particles are protected from further reactions and aggregations, which increase their stability and longevity.

Starch: Hydroxypropyl starch obtained from monosaccharides (glucose and galactose) and disaccharides (maltose and lactose) can be used as both reducing and stabilizing agents to synthesize AgNPs with a relatively narrow size distribution (average particle size of 25 nm). Carbohydrate-based AgNPs ranged from 6 to 13 nm. They were highly effective against E. coli, S. aureus, B. subtilis, P. aeruginosa, C. albicans, and some gram-negative bacteria when used in water disinfection [27].

Honey: A very efficient method to synthesize AgNPs in the presence of honey as a reducing and stabilizing agent is also illustrated [28]. This process is pH controlled. At room temperature, highly crystalline nanoparticles of various sizes are formed. At pH 8.5, predominantly monodisperse, nearly spherical colloidal AgNPs with a size of 4 nm can be prepared. They are also found to show high bactericidal activities in water treatment technologies.

Tea Extract: Caffeine and theophylline in tea extracts are utilized for catalysis and synthesizing NPs. Ag NPs utilizing polyphenols present in tea extract and epicatechin were synthesized. Depending upon the ratio of water to tea extract used for the synthesis AgNPs sizes ranging from 4 nm to 100 nm are formed, and from 15 to 26 nm are formed when various ratios of epicatechin are used [29].

Vitamin B2: For the synthesis of Ag and PdNPs, vitamin B2 is an effective reducing and capping agent. It is an agent that is biodegradable and has low toxicity. Because of high solubility in water, it forms rod-like sNPs when water is used as the solvent. In other solvents such as acetic acid the particle sizes of Ag and Pd NPs are 4.0 nm - 6.0 nm, in ethylene glycol, 4.5 nm - 6.5 nm, and in N-methylpyrrolidinone (NMP), 5.0 - 6.0 nm sized particles are formed [30].

Green tea (Camellia sinensis): Green tea extract containing polyphenols was used to synthesize and stabilize colloidal Ag and AuNPs. Ag and AuNPs of average size 40 nm can be formed using highly efficient single photon-induced luminescence by carefully changing the concentrations of metal ions and the quantity of tea extract [31].

Banana peel (Musa paradisiaca): Banana peel extract is used as a new source for synthesizing AgNPs. The produced AgNPs displayed antifungal activity against C. albicans and C. lipolytica and antibacterial activity against E. coli, Shigella sp., Klebsiella sp. and E. aerogenes [32].

Orange (Citrus sinensis): Orange peel is also used as a reducing and stabilizing agent to prepare Ag NPs that are effective against *E. Coli, Ps. Aeruginosa* (gram-negative), and S. Aureus (gram-positive) [33].

Swietenia mahogany: Dried leaves of mahogany trees containing various polyhydroxy limonoids were used to synthesize Ag, Au, and Ag/Au alloy nanoparticles [34].

Cashew nut (Anacardium occidentale): Leaf extract of anacardium occidentale contains polyols that act as reducing and stabilizing agents to synthesize Au, Ag, Au-Ag alloy, and Au core-Ag shell NPs at room temperature. According to the XRD results, these NPs have highly crystalline and cubic structures [35].

Leafy milk hedge (Euphorbia nivulia): Stem latex of Euphorbia nivulia contains peptides and terpenoids used to synthesize Ag and CuNPs of an average diameter of 5 - 10 nm. These synthesized NPs are found very effective against gram-negative and gram-positive bacteria [36].

Jowar (Sorghum): Sorghum contains high levels of diverse phenolic compounds that can function as reducing and capping agents in the synthesis of uniform and spherical Fe and AgNPs of an average diameter of 10 to 50 nm at room temperature [37].

5.5 MICROBES FOR THE SYNTHESIS OF METAL NPs

Fungus Trichoderma viride: It can be used to synthesize Ag NPs (sizes ranging from 5 to 40 nm) at room temperature. They are found to be highly effective against various bacterial strains such as *S. typhi* (gram-negative rods), *E. coli* (gram-negative rods), *S. aureus* (gram-positive cocci), and *Micrococcus luteus* (gram-positive cocci) during water treatment methods [38].

Fungus Fusarium oxysporum: Rice husk with fungus Fusarium oxysporum can transform amorphous plant biosilica into crystalline silica at room temperature. It further forms spherical and highly crystalline SiO_2 NPs of average size 2 - 6 nm [39].

Bacteria and actinomycetes Thermomonospora specie: Prokaryotic microorganism Thermomonospora specie can be used to synthesize well-dispersed AuNPs of an average size 8 nm [40].

Bacterium Brevibacterium casei: It can synthesize Ag (sizes from 10 to 50 nm) and Au (sizes from 10 to 50 nm) NPs [41].

Lactobacillus sp. and Sachharomyces cerevisae: They can be used to synthesize TiO_2 NPs with sizes ranging from 8 to 35 nm. Titania NPs are very effective against pathogens found in water [42].

5.6 BIOPOLYMER-BASED NANOMATERIALS

Biopolymers are naturally occurring biodegradable polymers that are abundant in occurrence and have good mechanical strength. They have been found very efficient when used in water technologies [43]. Water filters are highly efficient in removing microorganisms such as bacteria and viruses from water. They are highly efficient, work at a higher rate, and have a larger surface area than traditional water filters. Peptides, chitosan, cellulose, chitin, derivatives, and the like, also have higher antimicrobial activity. They can be used as natural nanomaterial disinfectants with a low cost of production and technology. The natural peptides can be engineered into nanoscale structures of desired nano size, shape, and function. Then they are punctured by antimicrobial cells and create a path in their cell membranes. Thus because of the remarkable antimicrobial activities of biopolymer-based NPs against various microorganisms ranging from bacteria, viruses, and fungi, they have been extensively explored as a novel substitute for synthetic polymer-based adsorbents in water treatment [44]. These biopolymer-based disinfectants are environmentally friendly, and their advanced strength, reactivity, and practicability of their variation characterizes them. They show tremendous selectivity towards organic compounds as they have reactive functional groups such as carboxyl, hydroxyl, amino, or acetamido groups. So, complete decontamination of impurities from water biopolymers based nanoadsorbents in water is highly efficient.

Nanocellulose-based nanomaterials: Nanocellulose is a renewable material. It is chemically non-reactive and shows good surface chemistry with hydrophilic behaviour. It has high mechanical strength and possesses a high surface area. It can be used as a membrane in nano filters in water treatment technologies. It has high potential and works well in purifying polluted water to remove microorganisms such as bacteria. Cellulose nanocrystals and nanofibrils are commonly used nano cellulose materials prepared by catalysis and hydrolysis of nanocellulose. They are rod-like structures with dimensions ranging from 100 to 2000 nm in size and 2 to 20 nm in diameter. They have both types of surface groups, ionic and non-ionic, that selectively absorb contaminants from water. Membranes designed with nanocellulose have microstructures with very fine pores which selectively pass certain species while others are stopped. Cellulose nanocrystals are prone to flocculation as they have a very low surface charge density. Hydrophobicity and controlled surface chemistry of nanocellulose-based membranes successfully resolves potential issues with conventional membrane technology, both biofouling and organic.

Chitosan-based nanomaterials: Chitosan is a naturally occurring nanomaterial with excellent antimicrobial activities. It is made from arthropod shells, fungi, algae, and crustaceans. They are non-toxic materials formed by the deacetylation of chitin. Chitosan resembles cellulose with a β-(1–4) glycoside bonded 2-amino-2-deoxy-d-glucose monomer. It is composed of biomolecules such as proteins, lipids, and the like, which can undergo polymerization and can work under a specific pH range. In water storage tanks, chitosan has been used as a disinfectant as membranes, sponges, or surface coatings. It has many benefits compared to other disinfectants because it has higher antibacterial activity, namely, it shows higher activity against various bacteria, viruses, and fungi. Antibacterial properties of chitosan depend on many factors, as shown in Figure 5.4. Above all, it is non-toxic

Eco-friendly Nanomaterials

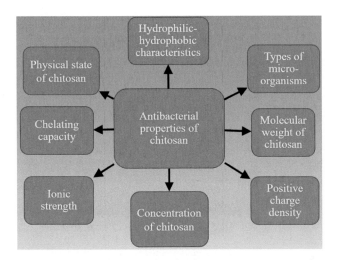

FIGURE 5.4 Factors affecting antibacterial properties of chitosan.

to humans. In water treatment methods, chitosan-coated membranes are used in a flow-through the membrane filtration system.

5.7 THE MECHANISM OF DISINFECTION BY ECOFRIENDLY NANOMATERIALS

Natural occurring nanomaterials show good antibacterial activities. Plant-based NPs, peptides, chitosan, and cellulose have been reviewed comprehensively for their antimicrobial action. These materials require little cost for production and also the process of disinfection is very simple. This is the reason why they are a good natural nanomaterial disinfectant.

5.7.1 Disinfection by Plant-Based NPs

Metal NPs such as nano-silver, nano-gold, nano-palladium, nano-aluminum, nano nickel, and the like, undergo strong interaction with the cell surface and pass through the cell membrane. These NPs have a high surface area to interact with antibodies and specifically inhibit the growth of infected cells without affecting normal cells. They generate reactive oxygen species (ROS) that cleaves DNA and RNA and inactivates microorganisms such as viruses and bacteria. The detailed mechanism of disinfection by NPs is shown in Figure 5.5.

Silver Nanoparticles: Although metal NPs such as iron-based NPs have exhibited good catalytic properties for organic reactions, silver has shown excellent performance as a disinfectant [45]. AgNPs show different pathways for their antimicrobial activities. They bind with enzymes through thiol functional groups and deactivate them by preventing DNA replication. The presence of

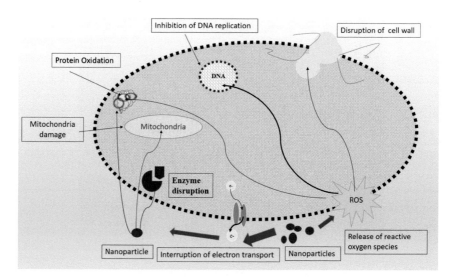

FIGURE 5.5 Schematic presentation showing the interaction of NPs and bacterial cells which includes the interaction of NPs with membrane, protein, and DNA.

AgNPs leads to the breakage of the cell membrane at endocytosis and other parts, through which NPs or ions penetrate the cells. Subsequently, AgNPs react with other species in the cell matrices, such as proteins or glutathione, to generate ROS or hydroxyl radicals via Fenton reactions. The excess of generated ROS eventually leads to cell death through the destruction of several species inside the cell matrix. Ahmed and coworkers synthesized AgNPs with biomolecules such as amino acids (alanine) and employed them for the disinfection of water. The synthesized Ala-Ag nanospheres were exploited for antibacterial behavior through disk diffusion antibacterial assay against gram-negative (Staphylococcus aureus) and gram-positive (Escherichia coli). Figure 5.6 shows the disk diffusion assay of these bacteria at different concentrations of 50 µg/mL, 100 µg/mL, and 150 µg/mL [46].

> **ZnO NPs:** ZnO NPs show extensive antibacterial due to their small particle size and high concentration. Their antimicrobial activity mechanism includes ROS production such as peroxide and hydroxyl radicals that damage the cell membrane and cytoplasm by depositing on them and deactivating cell activities.
>
> **TiO_2 NPs:** TiO_2 NPs inactivate microorganisms by the formation of ROS (OH• and peroxide radicals) between 300 and 390 nm wavelength irradiation. Microorganism inactivation by TiO_2 depends on certain factors such as nanoparticle concentration, type of microbe, the wavelength of radiation and its intensity, O_2 concentration, pH, ROS, and the retention time of active species.

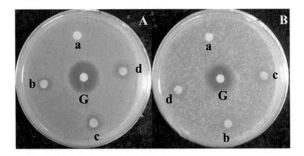

FIGURE 5.6 Disk diffusion assay of (A) *E. coli* and (B) *S. aureus* showing zone of inhibition in the presence of Ala–AgNPs at concentrations of (b) 50 µg/mL, (c) 100 µg/mL, and (d) 150 µg/mL. The controls represent (G) gentamycin and (a) silver nitrate substrates.

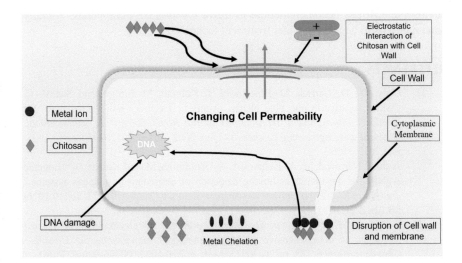

FIGURE 5.7 Schematic representation of antimicrobial mechanisms of chitosan.

Chitosan nanoparticles: The antimicrobial action of chitosan follows a specific mechanism [30]. At first, chitosan particles (positively charged) and microbial cell membrane (negatively charged) attract each other by electrostatic attractions, increasing the penetrability of membranes and outflow of cell organelles. This mechanism is applicable below pH 6. Similarly, chitosan derivatives containing quaternary ammonium groups, such as N, N, N-trimethyl chitosan, N-propyl-N, N-dimethyl chitosan, and N-furfuryl-N, N-dimethyl chitosan have high potential as antimicrobial agents compared to neutral chitosan. It is because of its encapsulation effect on microorganisms. It is also suggested that molecules cover the microbial cell surfaces and form an impervious wall around them, because of this, the cell ruptures and leaks its intracellular constituents and this causes the death of the microbial cell. But as pH rises (>6), chitosan or such other NPs lose their protonation tendency and become less effective [47] After that, chitosan chelates with trace metals and inactivates enzymatic activities of the microbial cells, as shown in Figure 5.7.

5.8 CONCLUSION

In recent years, studies on the depletion of water resources and deterioration of water quality by contaminants have been undertaken. It is clear that the need for good quality water has emerged for a secure future for human life. Water has no other alternatives, and it cannot be replaced with anything else. There are many investments made in water purification, and it seems now that nanotechnological tools can be helpful in solving current water problems. In this context and after knowing the harmful effects of engineered NPs on human health and the environment, it has become essential to move towards environmentally friendly nanomaterials to be used in water treatment technology.

REFERENCES

[1] Grey, D., D. Garrick, D. Blackmore, J. Kelman, M. Muller, and Sadoff, C. "Water security in one blue planet: twenty-first century policy challenges for science." *Philosophical Transactions of the Royal Society A: Mathematical, Physical and Engineering Sciences* 371, no. 2002 (2013), 20120406. doi:10.1098/rsta.2012.0406.

[2] Joshi, P. Nemiwal, M. Al-Kahtani, A. A. Ubaidullah, M. and Kumar, D. "Biogenic AgNPs for the non-cross-linking detection of aluminum in aqueous systems." *Journal of King Saud University—Science* 33, no. 6 (2021), 101527. doi:10.1016/j.jksus.2021.101527.

[3] Nemiwal, M. and Kumar, D. "Recent progress on electrochemical sensing strategies as comprehensive point-care method." *Monatshefte für Chemie—Chemical Monthly* 152, no. 1 (2021), 1–18. doi:10.1007/s00706-020-02732-0.

[4] Pond, K. R. Cronin, A. A. and Pedley, S. "Recreational water quality in the Caspian Sea." *Journal of Water and Health* 3, no. 2 (2005), 129–138. doi:10.2166/wh.2005.0013.

[5] Bridle, H. "Nanotechnology for detection of waterborne pathogens." *Waterborne Pathogens* (2021), 293–326. doi:10.1016/b978-0-444-64319-3.00009-5.

[6] Ojha, A. "Nanomaterials for removal of waterborne pathogens." *Waterborne Pathogens* (2020), 385–432. doi:10.1016/b978-0-12-818783-8.00019-0.

[7] Bali, U. "Ferrioxalate-mediated photodegradation and mineralization of 4-Chlorophenol." *Environmental Science and Pollution Research* 10, no. 1 (2003), 33–38. doi:10.1065/espr2002.10.134.

[8] Nemiwal, M. Zhang, T. C. and Kumar, D. "Recent progress in g-C3N4, TiO2 and ZnO based photocatalysts for dye degradation: strategies to improve photocatalytic activity." *Science of the Total Environment* 767 (2021), 144896. doi:10.1016/j.scitotenv.2020.144896.

[9] Nemiwal, M. and Kumar, D. "TiO2 and SiO2 encapsulated metal nanoparticles: synthetic strategies, properties, and photocatalytic applications."

[10] Jindal, H. Kumar, D. Sillanpaa, M. and Nemiwal, M. "Current progress in polymeric graphitic carbon nitride-based photocatalysts for dye degradation." *Inorganic Chemistry Communications* 131 (2021), 108786. doi:10.1016/j.inoche.2021.108786.

Inorganic Chemistry Communications 128 (2021), 108602. doi:10.1016/j.inoche.2021.108602.

[11] Kumari, S. Sharma, K. S. Nemiwal, M. Khan, S. and Kumar, D. "Simultaneous detection of aqueous aluminum(III) and chromium(III) using *Persea americana* reduced and capped silver nanoparticles." *International Journal of Phytoremediation*, 2021, 1–14. doi:10.1080/15226514.2021.1977911.

[12] Chaturvedi, S. Dave, P. N. and Shah, N.K. "Applications of nano-catalyst in new era." *Journal of Saudi Chemical Society* 16, no. 3 (2012), 307–325. doi:10.1016/j.jscs.2011.01.015.

[13] Nemiwal, M. Gosu, V. Zhang, T. C. and Kumar, D. "Metal organic frameworks as electrocatalysts: Hydrogen evolution reactions and overall water splitting." *International Journal of Hydrogen Energy* 46, no. 17 (2021), 10216–10238. doi:10.1016/j.ijhydene.2020.12.146.

[14] Nemiwal, M. Zhang, T. C. and Kumar, D. "Graphene-based electrocatalysts: Hydrogen evolution reactions and overall water splitting." *International Journal of Hydrogen Energy* 46, no. 41 (2021), 21401–21418. doi:10.1016/j.ijhydene.2021.04.008.

[15] Nemiwal, M. Sharma, V. and Kumar, D. "Improved designs of multifunctional covalent-organic frameworks: hydrogen storage, methane storage, and water harvesting." *Mini-Reviews in Organic Chemistry* 18, no. 8 (2021), 1026–1036. doi:10.2174/1570193x17999201127105752.

[16] Nemiwal, M. and Kumar, D. "Metal organic frameworks as water harvester from air: hydrolytic stability and adsorption isotherms." *Inorganic Chemistry Communications* 122 (2020), 108279. doi:10.1016/j.inoche.2020.108279.

[17] Nemiwal, M. Gosu, V. Dhillon, A. and Kumar, D. "Environmental applications of metal-organic frameworks: recent advances and challenges." In *Metal−Organic Frameworks for Environmental Remediation*, ACS Symposium Series (2021), Editors: Smita S. Kumar, Pooja Ghosh and Lakhveer Singh, 299–318, https://doi.org/10.1021/bk-2021-1394.ch012.

[18] Singh N. Dhilon, A. Nemiwal, M. and Kumar, D. "Metal−organic frameworks for water decontamination and reuse: a dig at heavy metal ions and organic toxins." In *Metal−Organic Frameworks for Environmental Remediation*, ACS Symposium Series (2021), Editors: Smita S. Kumar, Pooja Ghosh and Lakhveer Singh, 77–124. doi:10.1021/bk-2021-1395.ch004

[19] Ahluwalia, V. Kumar, J. Sisodia, R. Shakil, N. A. and Walia, S. "Green synthesis of silver nanoparticles by Trichoderma harzianum and their bio-efficacy evaluation against Staphylococcus aureus and Klebsiella pneumonia." *Industrial Crops and Products* 55 (2014), 202–206. doi:10.1016/j.indcrop.2014.01.026.

[20] Kanchi, S. "Nanotechnology for Water Treatment." *Journal of Environmental Analytical Chemistry* 01, no. 02 (2014). doi:10.4172/2380-2391.1000e102.

[21] Sharma, Y.C. Srivastava, V. Singh, V.K. Kaul, S.N. and Weng, C.H. "Nano-adsorbents for the removal of metallic pollutants from water and wastewater." *Environmental Technology* 30, no. 6 (2009), 583–609. doi:10.1080/09593330902838080.

[22] Wang, L. Hu, C. and Shao, L. "The antimicrobial activity of nanoparticles: present situation and prospects for the future." *International Journal of Nanomedicine* 12 (2017), 1227–1249. doi:10.2147/ijn.s121956.

[23] Liu, Y. Xu, Z. and Li, X. "Cytotoxicity of titanium dioxide nanoparticles in rat neuroglia cells." *Brain Injury* 27, no. 7–8 (2013), 934–939. doi:10.3109/02699052.2013.793401.

[24] Al-Issai, L. Elshorbagy, W. Maraqa, M. Hamouda, M. and Soliman, A. "Use of nanoparticles for the disinfection of desalinated water." *Water* 11, no. 3 (2019), 559. https://doi.org/10.3390/w11030559

[25] Kumar, P. Tomar, V. Joshi, R. K. and Nemiwal, M. "Nanocatalyzed synthetic approach for quinazoline and quinazolinone derivatives: a review (2015–present)." *Synthetic Communications* (2022), (52),1–32. doi:10.1080/00397911.2022.2041667.

[26] Gardea-Torresdey, J. L. Gomez, E. Peralta-Videa, J. R. Parsons, J. G. Troiani, H. and Jose-Yacaman, M. "Alfalfa sprouts: a natural source for the synthesis of silver nanoparticles." *Langmuir* 19, no. 4 (2003), 1357–1361. doi:10.1021/la020835i.

[27] El-Rafie, M.H. El-Naggar, M.E. Ramadan, M.A. Fouda, M. M. Al-Deyab, S. S. and Hebeish, A. "Environmental synthesis of silver nanoparticles using hydroxypropyl starch and their characterization." *Carbohydrate Polymers* 86, no. 2 (2011), 630–635. doi:10.1016/j.carbpol.2011.04.088.

[28] Philip, D. "Honey mediated green synthesis of silver nanoparticles." *Spectrochimica Acta Part A: Molecular and Biomolecular Spectroscopy* 75, no. 3 (2010), 1078–1081. doi:10.1016/j.saa.2009.12.058.

[29] Moulton, M. C. Braydich-Stolle, L. K. Nadagouda, M. N. Kunzelman, S. Hussain, S. M. and Varma, R. S. "Synthesis, characterization and biocompatibility of 'green' synthesized silver nanoparticles using tea polyphenols." *Nanoscale* 2, no. 5 (2010), 763. doi:10.1039/c0nr00046a.

[30] Nadagouda, M. N. and Varma, R. S. "Green and controlled synthesis of gold and platinum nanomaterials using vitamin B2: density-assisted self-assembly of nanospheres, wires and rods." *Green Chemistry* 8, no. 6 (2006), 516. doi:10.1039/b601271j.

[31] Vilchis-Nestor, A. R. Sánchez-Mendieta, V. Camacho-López, M. A. Gómez-Espinosa, R. M. Camacho-López, M. A. and Arenas-Alatorre, J. A. "Solventless synthesis and optical properties of Au and Ag nanoparticles using Camellia sinensis extract." *Materials Letters* 62, no. 17–18 (2008), 3103–3105. doi:10.1016/j.matlet.2008.01.138.

[32] Deena, S. Dakshinamurthy, A. and Selvakumar, P. M. "Green synthesis of silver nanoparticle using banana sap." *Advanced Materials Research* 1086 (2015), 7–10. doi:10.4028/www.scientific.net/amr.1086.7.

[33] Kaviya, S., J. Santhanalakshmi, B. Viswanathan, J. Muthumary, and Srinivasan, K. "Biosynthesis of silver nanoparticles using citrus sinensis peel extract and its antibacterial activity." *Spectrochimica Acta Part A: Molecular and Biomolecular Spectroscopy* 79, no. 3 (2011), 594–598. doi:10.1016/j.saa.2011.03.040.

[34] Mondal, S. Roy, N. Laskar, R. A. Sk, I. Basu, S. Mandal, D. and Begum, N. A. "Biogenic synthesis of Ag, Au and bimetallic Au/Ag alloy nanoparticles using

aqueous extract of mahogany (Swietenia mahogani JACQ.) leaves." *Colloids and Surfaces B: Biointerfaces* 82, no. 2 (2011), 497–504.

[35] Sheny, D.S. Mathew, J. and Philip, D. "Phytosynthesis of Au, Ag and Au–Ag bimetallic nanoparticles using aqueous extract and dried leaf of Anacardium occidentale." *Spectrochimica Acta Part A: Molecular and Biomolecular Spectroscopy* 79, no. 1 (2011), 254–262. doi:10.1016/j.saa.2011.02.051.

[36] Valodkar, M. Nagar, P. S. Jadeja, R. N. Thounaojam, M. C. Devkar, R. V. and Thakore, S. "Euphorbiaceae latex induced green synthesis of non-cytotoxic metallic nanoparticle solutions: a rational approach to antimicrobial applications." *Colloids and Surfaces A: Physicochemical and Engineering Aspects* 384, no. 1–3 (2011), 337–344. doi:10.1016/j.colsurfa.2011.04.015.

[37] Njagi, E. C. Huang, H. Stafford, L. Genuino, H. Galindo, H. M. Collins, J. B. Hoag, G. E. and Suib, S. L. "Biosynthesis of Iron and Silver Nanoparticles at Room Temperature Using Aqueous Sorghum Bran Extracts." *Langmuir* 27, no. 1 (2010), 264–271. doi:10.1021/la103190n.

[38] Fayaz, A. M. Balaji, K. Girilal, M. Yadav, R. Kalaichelvan, P. T. and Venketesan, R. "Biogenic synthesis of silver nanoparticles and their synergistic effect with antibiotics: a study against gram-positive and gram-negative bacteria." *Nanomedicine: Nanotechnology, Biology and Medicine* 6, no. 1 (2010), 103–109. doi:10.1016/j.nano.2009.04.006.

[39] Bansal, V. Ahmad, A. and Sastry, M. "Fungus-mediated biotransformation of amorphous silica in rice husk to nanocrystalline silica." *Journal of the American Chemical Society* 128, no. 43 (2006), 14059–14066. doi:10.1021/ja062113+.

[40] Ahmad, A. Senapati, S. Khan, M. I. Kumar, R. and Sastry, M. "Extracellular biosynthesis of monodisperse gold nanoparticles by a novel extremophilic actinomycete, *thermomonospora* sp." *Langmuir* 19, no. 8 (2003), 3550–3553. doi:10.1021/la0267721.

[41] Kalishwaralal, K. Venkataraman Deepak, SureshBabu Ram Kumar Pandian, Muniasamy Kottaisamy, Selvaraj BarathManiKanth, Bose Kartikeyan, and Sangiliyandi Gurunathan. "Biosynthesis of silver and gold nanoparticles using Brevibacterium casei." *Colloids and Surfaces B: Biointerfaces* 77, no. 2 (2010), 257–262. doi:10.1016/j.colsurfb.2010.02.007.

[42] Prasad, K. Jha, A. K. and Kulkarni, A. R. "Lactobacillus assisted synthesis of titanium nanoparticles." *Nanoscale Research Letters* 2, no. 5 (2007), 248–250. doi:10.1007/s11671-007-9060-x.

[43] Altintas, Z. Gittens, M. Guerreiro, A. Thompson, K-A. Walker, J. Piletsky, S. and Tothill, I. E. "Detection of waterborne viruses using high affinity molecularly imprinted polymers." *Analytical Chemistry* 87, no. 13 (2015), 6801–6807. doi:10.1021/acs.analchem.5b00989.

[44] Botes, M. and Cloete, T. E. "The potential of nanofibers and nanobiocides in water purification." *Critical Reviews in Microbiology* 36, no. 1 (2010), 68–81. doi:10.3109/10408410903397332.

[45] Kumar, P. Tomar, V. Joshi, R. K. and Nemiwal, M. "Nanocatalyzed synthetic approach for quinazoline and quinazolinone derivatives: a review (2015–present)." *Synthetic Communications* (2022), (52), 1–32. doi:10.1080/00397911.2022.2041667.

[46] Naaz, F. Farooq, U. Khan, M. A. and Ahmad, T. "Multifunctional efficacy of environmentally benign silver nanospheres for organic transformation,

photocatalysis, and water remediation." *ACS Omega* 5, no. 40 (2020), 26063–26076.

[47] Qi, L. Xu, Z. Jiang, X. Hu, C. and Zou, X. "Preparation and antibacterial activity of chitosan nanoparticles." *Carbohydrate Research* 339, no. 16 (2004), 2693–2700. doi:10.1016/j.carres.2004.09.007.

6 Advance Nanomaterials-Based Adsorbents for Removal and Recovery of Metals from Water

W. M. Dimuthu Nilmini Wijeyaratne
Department of Zoology and Environmental Management,
Faculty of Science, University of Kelaniya, Dalugama,
Kelaniya, Sri Lanka

CONTENTS

6.1	Heavy Metals: Sources and Environmental Effects	98
6.2	Nanomaterials	100
	6.2.1 Organic Nanomaterials	101
	6.2.2 Inorganic Nanomaterials	101
6.3	Nanomaterial Based Environmental Remediation Technologies	102
6.4	Nanofiltration	102
6.5	Nanomaterial Based Coagulation/Flocculation	103
6.6	Nanomaterial Based Adsorption	103
6.7	Diverse Types of Nanoadsorbents Used in Heavy Metal Remediation	104
	6.7.1 Carbon-Based Nano Adsorbents	104
	6.7.2 Metal-Based and Metal Oxide-Based Nanoadsorbents	107
	6.7.3 Silica-Based Nano Adsorbents	108
	6.7.4 Nanocomposite Adsorbents	108
6.8	Mechanisms of Adsorption	113
6.9	Environmental Factors Affecting Heavy Metal Removal from Nanomaterial	114
	6.9.1 Effect of the pH of the Medium	114
	6.9.2 Effect of Temperature of the Medium	115
	6.9.3 Effect of the Presence of Impurities	115
	6.9.4 Effect of Membrane Roughness	115

DOI: 10.1201/9781003252931-6

6.9.5 Effect of the Concentration/Dosage of the Adsorbent...................115
6.9.6 The Effect of the Contact Time 116
6.10 Future Perspectives in Nanomaterial-Based Heavy
Metal Adsorption...116
References...116

6.1 HEAVY METALS: SOURCES AND ENVIRONMENTAL EFFECTS

Heavy metals are metallic elements with a high atomic weight and a density of at least five times greater than water [1]. Heavy metals include transition metals, rare earth metals, and lead group elements [2]. Most of these heavy metals are natural substances that occur in relatively low concentrations in the environment. However, the expansion of agricultural and industrial activities has increased the anthropogenic input of heavy metals into the natural environment. The anthropogenic sources of heavy metals are indicated in Figure 6.1.

Heavy metals have a long persistence time in the environment as they are not degradable by biological or chemical processes. Further, they suppress the biodegradation of organic compounds and increase the adverse impacts of the same organic contaminants. The occurrence of the heavy metals in the natural environment in unacceptable concentrations can cause significant changes to the physical and chemical properties of the receiving environment. Further, the heavy metals are noxious at very low concentrations, and they can cause serious health effects to biota because of bioaccumulation via the food chains in the ecosystems. Based on epidemiological and toxicological studies, mercury, cadmium, chromium, arsenic, and lead are categorized as probable human carcinogens by the United

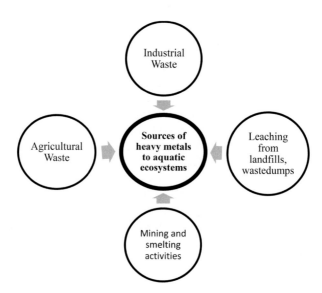

FIGURE 6.1 Potential anthropogenic sources of heavy metals to the aquatic environment.

TABLE 6.1
Potential Health Effects and the Sources of Common Heavy Metals as Described by USEPA [3]

Heavy Metal	Potential Health Effects from Long-Term Exposure	Potential Sources
Arsenic	Skin ulcers Circulatory system-related disorders High cancer risk	Erosion of As containing deposits Non-point source runoff from glass waste disposal sites Non-point source runoff from electronics waste production sites. Atmospheric deposition
Cadmium	Kidney damage High cancer risk	Erosion of Cd containing deposits. Non-point source runoff from metal refineries Non-point source runoff from batteries and paint related waste deposition sites Agricultural runoff
Chromium	Allergic skin reactions	Erosion of natural deposits Discharge from steel industries Discharge from pulp mills Agricultural runoff
Copper	Gastrointestinal tract related diseases Liver or Kidney damage	Erosion of natural deposits
Lead	Liver or Kidney damage High cancer risk	Erosion of Cd containing deposits Non-point source runoff from metal refineries, batteries and paint related waste deposition sites Agricultural runoff
Mercury	Kidney damage	Erosion of natural deposits Non-point source runoff from refineries and factories Agricultural runoff

States Environmental Protection Agency [1]. The harmful effects of heavy metals shown by USEPA are given in Table 6.1 [3]. Heavy metal contamination of the terrestrial and aquatic ecosystems has become a key environmental problem in recent decades.

Due to the detrimental environmental and health effects caused by the heavy metals in the natural environment there is a growing research concern around identifying possible remediation and recovery methods to protect the natural

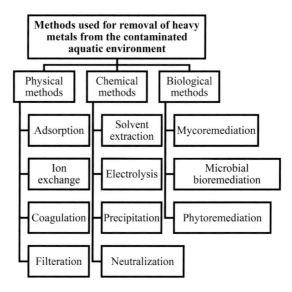

FIGURE 6.2 Physical, chemical, and biological methods of heavy metal removal.

ecosystems from heavy metal contamination. Several studies have identified successful biological, chemical, and physical methods to remove heavy metals from the contaminated environment. These methods are summarized in Figure 6.2.

Most of these methods are technology and cost intensive. Further, the physical and chemical methods are not environmentally friendly. There can be possible side effects that can threaten the balance of the ecosystems when these methods are utilized for heavy metal removal. The biological methods involved in heavy metal remediation use plant or microbial-based metabolism for heavy metal remediation and therefore take a longer period for the process, and sometimes these biological systems can be disturbed because of the toxicity of the heavy metals in the medium [4]. However, recent research has identified the possibility of using nanomaterial-based adsorbents as an ecofriendly and environmentally acceptable approach to the elimination of heavy metals from polluted environments. This chapter focuses on introducing the potential nanomaterial-based adsorbents that can remove heavy metals from contaminated aquatic ecosystems. Further, the removal mechanisms associated with these adsorbents and their effectiveness in remediation are described.

6.2 NANOMATERIALS

Nanomaterials are defined as materials that contain particles with a diameter ranging from 1.0 to 100 nm in at least a single dimension [5]. Nanomaterials are considered to be very useful in environmental remediation processes because of their unique properties that are helpful in the removal of potential pollutants.

The nanoscale particles which make up the nanomaterials have diverse surface interaction properties and fast adsorption rates, which increases the potential for removing contaminants from polluted environments. Further, the nanomaterials are highly reactive and have a large adsorption surface. They are tolerant to temperature modification and, therefore can be used under variable environmental conditions.

The nanomaterials can be mainly categorized into two major categories, organic nanomaterials and inorganic nanomaterials.

6.2.1 Organic Nanomaterials

The main constituent of organic nanomaterials is carbon. Carbon is one of the most commonly occurring elements in the cosmos and has unique strength and atomic arrangement properties. Carbon can have a wide variety of allotropes based on the atomic arrangement. Each of the allotropes has unique physical properties that are important in many industrial and environmental applications.

All nanomaterials consisting of carbon atoms are termed carbon-based or carbon nanomaterials. These carbon nano allotropes primarily consist of sp^2 carbon atoms arranged in a hexagonal network. Most the carbon-based nanomaterials have some common properties due to their similarity in structural arrangement. However, some significant differences in the properties have been recorded due to their different sizes and shapes. Carbon nanomaterials can be further classified according to their geometric structure [6].

As the name suggests, the NPs in carbon nanotubes have a tube shape. They comprise rolled-up sheets of single-layer carbon atoms formed a cylindrical nanotube (CWCNT). Several nanotubes can be interlinked concentrically to form MWCNT. The diameter of CWCNT is usually less than 1 nm, and the diameter of MWCNTs can reach up to 100 nm [7].

Carbon nanomaterials with horn-shaped particles are termed nanohorns. Carbon nanohorns (CNHs) contain conical-shaped carbon nanostructures arranged into tubules with a conical end. They are also termed carbon nanocones. Carbon nanohorns have high dispersibility, conductivity, and a large specific surface area compared to carbon nanotubes [7].

Fullerenes contain sphere or ellipsoid-shaped carbon nanomaterial arrangements in their structure. Fullerenes were the first carbon-based nanomaterials discovered in 1985 and are the smallest known stable carbon nanostructures. They represent the boundary between carbon molecules and carbon nanomaterials [8].

6.2.2 Inorganic Nanomaterials

Inorganic nanomaterials comprise of a metal element or non-metal element combined with other atoms to form a hydroxide, chalcogenide, oxide, or phosphate compound. Most common inorganic nanomaterials are made up of AuNPs, AgNPs, and Fe_2O_3 NPs. Inorganic NPs are highly photostable and can be visualized by luminescence imaging, fluorescence imaging, and X-ray computed tomography techniques. These properties of inorganic nanomaterials help them be widely applicable in the medical field for drug delivery processes and tissue engineering in cancer patients [9].

Applications of both organic and inorganic nanomaterials in heavy metal removal will be discussed in section 6.3.

6.3 NANOMATERIAL BASED ENVIRONMENTAL REMEDIATION TECHNOLOGIES

Environmental remediation is defined as the use of chemical and physical technologies such as chemical reactions, photocatalysis, absorption, filtration, and adsorption for the removal of contaminants from the soil, water, and air environments [10].

The pore size, thickness, surface characteristics, and morphology of a nanomaterial directly affect its contaminant removal capability. The higher surface-to-volume ratio of the nanomaterials enhances reactivity and increases the effectiveness of contaminant removal compared to techniques where non-nanomaterial is involved. The surface chemistry of nanomaterials can be changed by grafting functional groups. These functional groups can target specific pollutants of interest and thus enhance remediation efficiency.

In the recent decade, nanomaterial-based environmental remediation techniques have become a key research direction in aquatic environment improvement studies. Among the vast applications of nanomaterial-based environmental remediation methodologies, several nanomaterial-based heavy metal remediation approaches include adsorption, nanofiltration, and coagulation/ flocculation.

6.4 NANOFILTRATION

Nanofiltration membranes are thin membranes made from organic nanomaterial. The pore sizes of these membranes ranges from 0.1 to 10 nm. Their higher permeability, lower energy consumption, and higher flow rates have increased the applicability of this technique in various industries including in environmental remediation. The nanofiltration technique is seen as a very reliable surface water treatment technique and is mainly used for removing excess salt and divalent metal ions from groundwater and surface water [11]. The nanofiltration technique is used for the elimination of pathogenic microorganisms and viruses from contaminated water.

Further, nanofiltration is widely used to reduce total dissolved solids, hardness, color, and odors from wastewater. Most of these studies were tested in laboratory conditions and were applicable to pilot-scale studies. However, recent advancements in nanofiltration-based techniques have resulted in the availability of a wide range of nanofiltration membranes that can be successfully implemented to remove heavy metals and pesticides from contaminated surface water and groundwater. Further advanced nanofiltration techniques are used to separate organic compounds in wastewater treatment and water recycling in industrial processes [12].

However, due to the high operational and maintenance costs associated with the nanofiltration process, it is not commonly used in commercial-scale water treatment. Further, metal oxides, colloids, and microorganisms can cause blockage of the pores and membrane fouling, which can reduce the efficiency of the nanofiltration procedure [13]. Therefore, it is vital to pretreat the permeate and control the pH, flow rate, and temperature of the permeate for proper maintenance and prevention of fouling and clogging of the nanofiltration membrane.

6.5 NANOMATERIAL BASED COAGULATION/FLOCCULATION

Flocculation is a physical environmental remediation technique. This technique causes a combination of small-sized pollutant particles to form larger aggregations of flocs. After the formation of the flocs or aggregations, they are removed by physical and chemical methods. There are two flocculation methods: direct and indirect flocculation [14]. In direct flocculation, chemical flocculants are not used, and the particles are flocculated by altering environmental parameters such as temperature and pH. In indirect flocculation, the chemical compounds are added to the medium to facilitate the formation of flocs. The direct flocculation method is relatively faster, and larger flocs are produced, whereas in the indirect flocculation, small fragile flocs are produced slowly.

There are four types of flocculation mechanisms for contaminant removal by nano flocculants [15]

- Charge neutralization: Adsorption of polyelectrolytes on oppositely charged sites of nano colloidal particles reduces net electrical charge.
- Electrostatic charge patching: Interactions occur between positively and negatively charged surfaces.
- Bridging: Assembly of particles with each other links them together.
- Sweep flocculation: Precipitates are used to remove colloidal contaminants.

Both natural and synthetic, organic and non-organic nanomaterial have been tested for use as flocculants in environmental remediation technologies. Many nanomaterials, including cellulose, chitosan, carbon nanotubes, and metal and metal oxide-based nanomaterials, have been identified as successful remediators in many environmental applications [16].

6.6 NANOMATERIAL BASED ADSORPTION

Adsorption is the removal of organic and inorganic matter by attaching it (adsorbate) to the surface of the adsorbing material (adsorbent) through physical and/or chemical forces such as van der Waals and electrostatic interactions. The formation of weak van der Waals forces between adsorbent and adsorbate is called physiosorption. Physiosorption is an endothermic process and a reversible reaction. In chemosorption, chemical reactions occur between the adsorbent and adsorbate. Chemisorption selectively binds adsorbates to the adsorbents, and it is an irreversible and exothermic process [17].

Adsorption is an environmentally friendly and cost-effective approach compared to other techniques used in environmental remediation. The removal of heavy metals by adsorption is very commonly practiced because of its high removal capacity, easy implementation, and the possibility of the easy recovery of adsorbed heavy metals by simple treatment. There are many organic and inorganic compounds used as adsorbents in environmental remediation. Both synthetic and natural adsorbent materials are used in environmental remediation applications [18]. Mainly activated carbon, mineral adsorbents including zeolite, silica, clay, chitosan and its derivatives, magnetic adsorbents, metal-organic composite adsorbents, and bio sorbents are used for the removal of heavy metals. The characteristics of an ideal adsorbent used for heavy metal removal include high sorption capacity, high selectivity, environmentally safeness, easy recyclability, and easy recoverability of the adsorbate [19]. Nanotechnological advancements and research based on the applications of nanomaterials have identified that nanomaterial-based adsorbents can satisfy most of these criteria and are considered one of the most effective and efficient types of adsorbents that can be practically applied for heavy metal removal from the contaminated water and soil ecosystems.

The nanomaterials used for adsorption include carbon-based and inorganic nanomaterial and are commonly termed nanoadsorbents. These different nanoadsorbents have unique properties that help in effective adsorption and recovery of selected heavy metals from the environment [20].

6.7 DIVERSE TYPES OF NANOADSORBENTS USED IN HEAVY METAL REMEDIATION

The Nanoadsorbents used in heavy metal remediation can be broadly categorized into four groups according to their composition. They are carbon-based nanoadsorbents, metal and metal oxide-based nanoadsorbents, Silicon-based nanoadsorbents, and nanocomposite adsorbents.

6.7.1 CARBON-BASED NANO ADSORBENTS

Carbon-based NPs are a very diverse group of nanomaterials in terms of the structure and the arrangement of the carbon atoms. They include carbon nanotubes, fullerenes, graphene, and derivatives of graphene such as carbon-based quantum dots, nano-diamonds, and graphene oxide. Carbon-based nanomaterials can be categorized into four categories according to their dimensions (Baby et al. 2019).

- 0D carbon-based nanomaterial: where the three dimensions of the nanomaterial (length, breadth and height) are less than 100 flocculated

by altering the environmental parameters, for example, fullerene and quantum dots.
- 1D carbon-based nanomaterial: where just one dimension of the nanomaterial is larger than 100 nm, and the other two are smaller than 100 nm, for example, carbon nanotubes.
- 2D carbon-based nanomaterial: where two dimensions of the nanomaterial are larger than 100 nm, and the other dimension is smaller than 100 nm, for example, graphene.
- 3D carbon-based nanomaterial: where all of the dimensions of the nanomaterial are larger than 100 nm, for example, graphite.

The structure of these carbon-based nanomaterials is given in Figure 6.3.

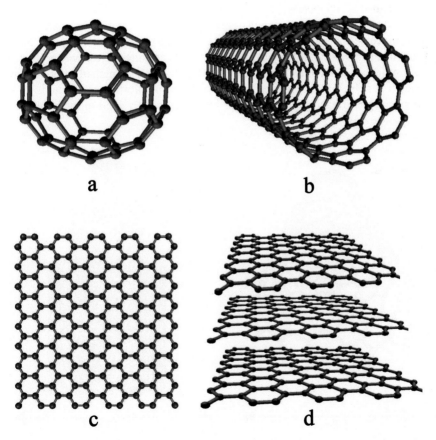

FIGURE 6.3 Structural differences of carbon-based nanomaterial (a: 0D structure; b: 1D structure; c: 2D structure; d: 3D structure) ([86]. License: http://creativecommons.org/licenses/by/4.0/).

Fullerenes have a cage-like atomic arrangement and are hydrophilic structures. The structural, chemical, and physical properties of fullerenes provide a large area of adsorption and have been used as adsorbents to eliminate heavy metals from contaminated aquatic environments. Several studies have shown that fullerenes tend to remove Cu ions better in comparison to the other carbon-based nano adsorbents [22].

Carbon nanotubes are a hollow tube-like arrangements of carbon atoms. The schematic representation of SWCNTs and MWCNTs is given in Figure 6.4.

The length of the nanotubes can vary from several nanometers to micrometers. Both SWCNTs and MWCNTs have high aspect ratios. These carbon nanotubes have excellent adsorbent properties as heavy metals can be adsorbed onto the surface of the nanotubes' walls and the interstitial spaces. Further, some heavy metals can be incorporated into the hollow tubules. The structure and chemical properties of the carbon nanotubes promote desorption, making it easier to recover the adsorbed heavy metals [6].

A 2D carbon nanomaterial such as graphene is widely used to eliminate organic contaminants from water. The structure of graphene comprises a hexagonal lattice made up of carbon atoms (Figure 6.3). The surface chemistry, electronic properties, high surface area, thermal mobility, and high mechanical strength of graphene have made it possible for it to be used in environmental purification applications [6–8]. Graphene-based nano adsorbents have been successfully used in the removal of

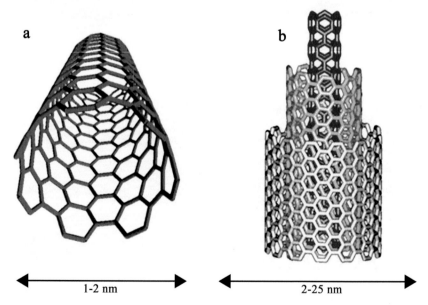

FIGURE 6.4 The schematic representation of (a) Single-walled (SWCNT) and (b) Multiwalled carbon nanotubes (MWCNT) ([86]. License: http://creativecommons.org/licenses/by/4.0/).

As, Pb, Cu, Ni, Cd, and Cr from contaminated aquatic ecosystems [23, 24]. Surface modification of graphene-based nanomaterials can further improve the adsorption efficiency. However, the graphene-based nano adsorbents convert into clusters and form aggregates which result in difficulties in the recovery of the adsorbed heavy metals [25].

Activated carbon is a three-dimensional carbon-based nano material that is most used in environmental remediation applications. Additionally, it is used in removing pollutants from aquatic and atmospheric media. However, it is most widely used as a recyclable and low-cost material for the decontamination of aquatic resources polluted with heavy metals, organic dyes, and other organic compounds. A wide range of research has been conducted in the past decade on different raw materials that can produce activated carbon-based nanomaterials. Organic materials such as chicken manure and agricultural sawdust waste are considered low-cost materials that can produce highly efficient heavy metal adsorbents. These activated carbon-based nanomaterials have shown efficient adsorbance towards heavy metal mixtures consisting of Fe, Pb, Cu, Zn, Ni, and Cd with a removal efficiency greater than 98% [26–28].

6.7.2 Metal-Based and Metal Oxide-Based Nanoadsorbents

Zerovalent iron nano particles (nZVI), AgNPs, and AuNPs are the most applied metal-based nano adsorbents used in environmental remediation studies. Zerovalent FeNPs comprise Fe(0) coated by a ferric oxide layer. The outer ferric oxide layer of the zerovalent FeNPs provides reaction sites with electrostatic interactions with the heavy metals in the medium [29]. The heavy metal elimination mechanism by nZVI involves redox, cementation, adsorption, and precipitation processes. Several studies have shown that nZVI is effective in the removal of Zn, Cd, Cu, Ni, and Pb [30–32].

Both AgNPs and AuNPs are considered superior NPs because of their unique chemical and physical characteristics. However, AgNPs are relatively unstable, low cost, and more abundant than AuNPs. Syntheses of AgNPs and AuNPs involves physical, chemical, and photocatalytic methods. Further, recent research has attempted green syntheses of AgNPs using biological objects such as plant extracts and secretions from microorganisms [33]. Green synthesized AgNPs have been effective in the removal of Pb, Co, Hg, and Cd [34–36], and AuNPs have been effective in the removal of Hg from contaminated water [37]. Further, enhanced visible surface plasmon resonance color changes, improved catalytic activity, and high chemical stability of AuNPs have made them applicable in the selective detection of various pollutants in aquatic environments and effective removal of those from the contaminated environment [38].

Metal oxide nanoparticles (MONPs) comprise of pure metal precursors. These metal precursors can produce many oxide compounds at different oxidation states. Metal oxides of alkali metals and noble metals such as Ag, Au, and Cu have a wide absorption capacity identifiable in the visible electromagnetic spectrum zone [39]. Nanosized metal oxides have a high heavy metal selectivity and exceptionally high heavy metal removal capacity. Some metal oxide-based nanomaterials include magnesium oxides, aluminum oxides, manganese oxides, titanium oxides, cerium oxides, nano-sized iron oxides, zinc oxides, and zirconium oxides. These metal oxide-based nano particles have unique properties that aid in effective adsorption, and they are abundant, nontoxic, and environmentally friendly materials, increasing their applicability in pollution remediation.

Among the metal oxide-based nanoadsorbents, the nanosized iron oxides, manganese oxides, and titanium oxides are widely applied in heavy metal removal from wastewater [40]. Iron-based nano oxides are the most diverse and most studied group of nanometal oxides used in heavy metal remediation. Iron oxides is a composite term used for oxides, hydroxides and oxy-hydroxides made up of Fe(II) and/or Fe(III) cations and O^{2-} and/or OH^- anions [41]. There are sixteen pure phases of nano iron oxides that have been widely used in water treatment. These sixteen pure phases are given in Figure 6.5.

Some applications of metal oxide-based nano particles in heavy metal remediation and their adsorbent capacities are summarized in Table 6.2.

The adsorbent capacity of metal oxide-based nanobsorbents is dependent on the temperature, pH, and pressure of the reaction medium. In addition, the wastewater flow rate can also influence the adsorbent capacity of MONPs. Usually, MONPs are impregnated with supporting molecules. These supporting molecules form composite nano adsorbents with a higher adsorption capacity [42].

6.7.3 Silica-Based Nano Adsorbents

Silicon nanoparticles are nontoxic and biocompatible nanomaterials comprising crystallized Si atoms crystallized arrangement. Different silicon-based nanomaterials include silicon nanosheets (SiNSs), silicon nanoparticles (SiNPs), and silicon nanotubes (SiNTs) [55]. Most silicon-based nanomaterials do not have specific surface properties, and as a result, they have limited applications to be used in ion exchange and adsorption. However, these silicon nanomaterials have been impregnated with various functional groups to improve their structure and functionality to be effectively used in environmental remediation applications. The different types of functional groups commonly used to improve the functionality of silicon nanomaterial are given in Table 6.3.

6.7.4 Nanocomposite Adsorbents

Nanocomposite adsorbents are also termed nanohybrid adsorbents. Composite nano adsorbents are modified materials that include combinations of carbon-based and metal-based NPs. The classification of nanocomposites according to the structure and raw material is given in Figure 6.6.

The two broad groups of nanocomposites include polymer-based and non-polymer-based nanocomposites (Figure 6.6). The nanocomposite adsorbents are lighter than the conventional NPs and have excellent thermal and mechanical

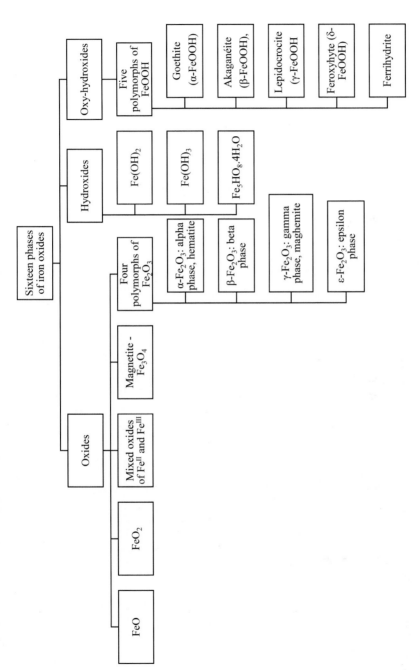

FIGURE 6.5 The sixteen pure phases of nano iron oxides used for water treatment applications.

TABLE 6.2
Application of Selected Nano Metal Oxides in Heavy Metal Remediation by Adsorption

Nanoadsorbent	Metals Removed	Adsorption Capacity (%)	Adsorption Conditions	References
Nano Iron Oxides	Cr (VI)	24.76%	Removal was higher in acidic pH	[43][44]
	As (V)	99%	Very high rate of adsorption in acidic pH	[45]
	Pb (II)	53%	Maximum adsorption at pH 5	[43]
	Cd (II)	91.6%	Maximum percentage removal of was achieved at alkaline medium (pH 9).	[46]
Nano Manganese oxides	Li	28.8%	High selectivity and efficiency in adsorbing lithium ion in seawater. Best adsorption efficiency was recorded from nano-$Li_{1.33}Mn_{1.67}O_4$ precursor compound formed at 500°C	[47]
	Hg(II)	80–95%	Optimum performance at pH 6 - 9, endothermic process. Spontaneous reaction. Increasing ionic strength reduced Hg(II) uptake.	[48]
	Pb	21.3	The optimum adsorption was observed an acidic medium at 60°C	[49]
	Fe	60.4	Adsorption was efficient and at low temperatures; exothermic reaction	[50]

Nano titanium oxides	Zn, Cd, Pb, Ni, Cu	90 - 100%		Optimum performance was at pH 8.	[51]
	As (III) and As (V)	50 - 65% for As (V) 25%–30% for As (III)		Adsorption capacity of As(V) was higher than As(III). The solution pH appeared to influence the arsenic adsorbing capacity.	[52]
	Pb (II), Cu (II), Fe (III), Cd (II) and Zn (II)	Pb: 97.06 Cu: 75.24 Fe: 79.77 Cd: 61.89 Zn: 35.18	Stronger adsorption affinity for Pb compared to Cd and Zn ions.		[53]
Nano magnesium oxides	Cd (II), Pb (II)			Stronger adsorption affinity for Cd than Pb ions.	[54]

TABLE 6.3
Functional Groups Used to Improve The Functionality Of Si-Based Nanomaterial

Functional Group	Loading Method	Functional Improvements	References
Metal Phosphate	Surfactant Templating, Layer-by-layer method, Wetness Impregnation	Negative charges of the surface hydro phosphate groups increase the ion exchange capability.	[56–59]
Molybdophosphate	Crystallization sol gel method, Direct impregnation	Stability in acidic environments. Improves the solvent distribution and complex formation	[60–62]
Titanate-Based Materials	Sol-gel method	Increases the adsorption efficacy of Sr^{2+} ions	[63–66]
Hydroxides and their modified forms	Direct incorporation into the Si matrix	Increases the number of exchangeable cations and increases the selective ion exchange capacity	[67–69]
Organic material	Direct incorporation into the Si matrix	Increases the adsorption capacity of Actinides and radioactive metals	[70–73]

properties. Further, adsorption efficiency is increased due to the high surface-to-volume ratio of nanocomposite material [74].

The functionality of the original nanomaterial can be increased by incorporating functional groups such as $-NH_2$ and $-SH$. Further, the use of polyethyleneimine, polypyrrole, humic acid, polyrhodanine, and MnO_2 as a coating for the iron/iron oxide NPs produces magnetic nanocomposites. Magnetic nanocomposites are very effective adsorbents in the remediation of heavy metals in the aquatic environment

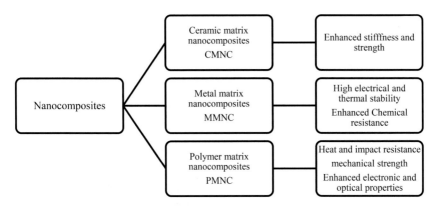

FIGURE 6.6 Classification of a nanocomposite material.

because of their high selectivity and high binding capability. The adsorbate can be easily recovered with a powerful magnetic field [75]. Magnetic nanocomposite-based heavy metal remediation has become increasingly popular because of the low energy requirements and associated costs entailed by this process [75].

The nanomaterials can be modified by incorporating them into naturally available macromolecules such as natural clays (zeolite, montmorillonite, kaolin) and dendrimers. Nano impregnated zeolite, montmorillonite, kaolin, and dendrimers have high functionality, higher cation exchange capacities, high sorption capacities, and high loading capacity for pollutant encapsulation by their nanocavities compared to the naturally occurring clay minerals [76].

6.8 MECHANISMS OF ADSORPTION

Several physical and chemical mechanisms are involved in adsorbing heavy metal ions onto nano adsorbents. These mechanisms are summarized in Figure 6.7. In the adsorptive removal of heavy metal ions, van der Waals forces adhere heavy metals to the surface of the nanoadsorbent. These forces can be easily reversed during desorption, and it helps for easy recovery of the heavy metal ions adsorbed onto the NPs.

Diffusion and active transport of heavy metals through the nano adsorbent membranes is another mechanism involved in environmental remediation. The small pores and the surface electric charges of nano adsorbent membranes remove charged heavy metal ions from the contaminated media. The nano adsorbent membranes reject charged solutes that are smaller than the membrane pores [77].

When the nanomaterials are functionally modified, the functional groups attached to the surface of the NPs can precipitate or make complexes with the heavy metals in the aqueous medium. In addition, some functional groups increase the cation exchange capacity of the nano adsorbents, and the heavy metals can be adsorbed by ion exchange.

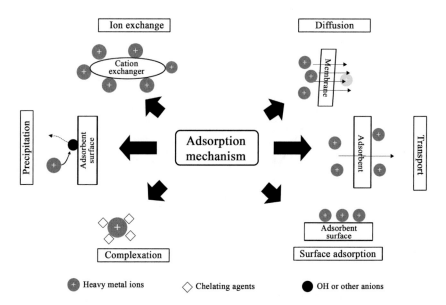

FIGURE 6.7 Types of mechanisms involved in heavy metal remediation. ([87]. License: http://creativecommons.org/licenses/by/4.0/).

The adsorption isotherms, adsorption kinetics, and thermodynamics explain heavy metal adsorption mechanisms by the nonadsorbent. Adsorption models can be developed based on these properties of adsorbent reactions. The concentration of adsorbents can further influence the adsorption reaction mechanism and adsorbent rate, functional groups, their functionality, the concentration of the adsorbates, the equilibrium parameters, and contact time.

6.9 ENVIRONMENTAL FACTORS AFFECTING HEAVY METAL REMOVAL FROM NANOMATERIAL

6.9.1 Effect of the pH of the Medium

The positive or negative charge of the surface of the nano adsorbents highly depends on the pH of the medium. Further, pH affects the bioavailability and speciation of heavy metals. At acidic conditions, conversion of H^+ ions to H_2 takes place at a higher rate as more protons are reacting with the nano surfaces. Therefore, at acidic pH, there can be more reactive H^+ and a faster reduction rate [78]. Further, it has been recorded that in natural pH media, electrostatic sorption, precipitation, and surface coordination are stronger than those in alkaline or acidic media.

Therefore, it has been recorded that most heavy metal sorption reactions are faster and stronger in acidic to neutral pH conditions than those in the basic medium [79].

6.9.2 Effect of Temperature of the Medium

Changes in the temperature can significantly affect the equilibrium of the adsorption reaction. Further, the energy required for adsorption reactions is determined by the medium's temperature [80]. The exothermic and endothermic reactions respond differently to the changes in temperature. When the exothermic reactions are considered, as the temperature increases, the active sites in the adsorbent decrease, and therefore there is a tendency to decrease the absorbance rate. In the endothermic reactions, the adsorbent surfaces are activated at higher temperatures. Therefore, the efficiency of the adsorption process of the endothermic reactions increases at higher temperatures [81], [80].

However, the performances of the natural material-based adsorbents are limited to an optimum temperature range. If the temperature is above or below the optimum range, the reactivity of the adsorbent may not be sufficient for the effective removal of heavy metals [81]. Further, the natural sorbent material can deteriorate the surface quality at extreme temperatures, resulting in disturbing or completely ceasing the heavy metal adsorption.

6.9.3 Effect of the Presence of Impurities

The small pores of the nano adsorbents can be clogged due to impurities in the medium. Therefore, the membranes may not perform the intended environmental remediation process. Further, if the clogging retains for a longer period, it can result in membrane fouling which can cause deterioration of the adsorbent.

6.9.4 Effect of Membrane Roughness

The roughness of the membrane is also important for proper adsorption. The roughness of the membranes can help impurities to adhere to the membrane surface for a lengthier period and reduce the efficiency of adsorption [82].

6.9.5 Effect of the Concentration/Dosage of the Adsorbent

The dosage/concentration of the nanomaterials used in the treatment is an important parameter that affects the efficiency and rate of adsorption. Increasing the adsorbent dosage is positively correlated with the adsorbing efficiency as there is more surface area available for the heavy metal ions to bind onto [83]. However, the adsorption rate gradually decreases if the adsorbent surface becomes saturated with heavy metal ions. Further, aggregation or agglomeration of heavy metals on the surface of the adsorbent can also reduce the surface area, resulting in a reduction of the absorbance rate [84].

6.9.6 THE EFFECT OF THE CONTACT TIME

The effect of the contact time on the rate of adsorption can vary throughout the adsorption process. It can show significant variations during the reaction's initial, intermediate, and final phases. The adsorption rate increases in the initial phase until the reaction achieves equilibrium [85]. After the equilibrium state, the adsorption rate decreases as heavy metals occupy most of the adsorption sites. The pseudo-first order can describe the effect of contact time on the rate of adsorbance, the pseudo-second order, the Zeldowitsch, and the Lagergren kinetic models [85].

6.10 FUTURE PERSPECTIVES IN NANOMATERIAL-BASED HEAVY METAL ADSORPTION

Research has shown that nanomaterial-based heavy metal adsorption is a very promising application in the environmental remediation of aquatic ecosystems. Different types, including carbon-based, silicon-based, metal-based, and composite nanomaterials, have been extensively studied and recommended for heavy metal remediation. However, most of these studies are conducted under controlled environmental conditions in the laboratory. Therefore, more research should be conducted to assess their performance in the natural environmental and climatic conditions. Further, the recovery of heavy metals from the adsorbents is also studied at a laboratory scale. The results have suggested that significant amounts of adsorbed heavy metals can be successfully recovered. However, when considering the variations in the natural environmental and climatic conditions in the tropical and temperate regions, the recovery percentages can show higher variations from the controlled laboratory conditions. Therefore, it is imperative to study the applicability of these studies in real-world situations. The cost and technological feasibility of applying nanomaterial-based heavy metal adsorption to the aquatic ecosystems also needs to be considered.

When considering the applications of nanomaterials in heavy metal remediation it is important to consider their environmental effects too. There is the consideration that nanomaterials themselves cause adverse effects when released into the natural environment. Therefore, more emphasis should be made on research directed towards eliminating the adverse environmental impact because of the release of nanomaterials and nanomaterial-based compounds used in the natural environment. Development of guidelines and policies that focus on responsible management and disposal of nanomaterial will be of timely importance for the future sustainability of the quality of the environment.

REFERENCES

[1] Tchounwou, P. B. Yedjou, C. G. Patlolla, A. K. and Sutton, D. J. "Heavy metal toxicity and the environment." *Experientia Supplementum* (2012), (101),133–164. doi:10.1007/978-3-7643-8340-4_6.

[2] Appenroth, K-J. "Definition of 'Heavy Metals' and Their Role in Biological Systems." *Soil Biology* (2009), (19), 19–29. doi:10.1007/978-3-642-02436-8_2.
[3] "National Primary Drinking Water Regulations | US EPA." https://epa.gov/ground-water-and-drinking-water/national-primary-drinking-water-regulations (accessed December 02, 2021).
[4] Akhtar, F. Z. Archana, K. M. Krishnaswamy, V. G. and Rajagopal, R. "Remediation of heavy metals (Cr, Zn) using physical, chemical and biological methods: a novel approach." *SN Applied Sciences* 2, no. 2 (2020). doi:10.1007/s42452-019-1918-x.
[5] Abbas, Q. Yousaf, B. Amina, Ali, M. U. Munir, M. A. El-Naggar, A. Rinklebe, J. and Naushad, M. "Transformation pathways and fate of engineered nanoparticles (ENPs) in distinct interactive environmental compartments: a review." *Environment International* 138 (2020), 105646. doi:10.1016/j.envint.2020.105646.
[6] Zaytseva, O. and Neumann, G. "Carbon nanomaterials: production, impact on plant development, agricultural and environmental applications." *Chemical and Biological Technologies in Agriculture* 3, no. 1, 1-26, (2016). doi:10.1186/s40538-016-0070-8.
[7] Utsumi, S. Miyawaki, J. Tanaka, H. Hattori, Y. Itoi, T. Ichikuni, N. Kanoh, H. Yudasaka, M. Iijima, S. and Kaneko, K. "Opening mechanism of internal nanoporosity of single-wall carbon nanohorn." *The Journal of Physical Chemistry B* 109, no. 30 (2005), 14319–14324. doi:10.1021/jp0512661.
[8] Georgakilas, V. Perman, J. A. Tucek, J. and Zboril, R. "broad family of carbon nanoallotropes: classification, chemistry, and applications of fullerenes, carbon dots, nanotubes, graphene, nanodiamonds, and combined superstructures." *Chemical Reviews* 115, no. 11 (2015), 4744–4822. doi:10.1021/cr500304f.
[9] Shen, J. Shafiq, M. Ma, M. and Chen, H. "Synthesis and surface engineering of inorganic nanomaterials based on microfluidic technology." *Nanomaterials* 10, no. 6 (2020), 1177. doi:10.3390/nano10061177.
[10] Guerra, F. Attia, M. Whitehead, D. and Alexis, F. "Nanotechnology for environmental remediation: materials and applications." *Molecules* 23, no. 7 (2018), 1760. doi:10.3390/molecules23071760.
[11] Mulyanti, R. and Susanto, H. "Wastewater treatment by nanofiltration membranes." *IOP Conference Series: Earth and Environmental Science* 142 (2018), 012017. doi:10.1088/1755-1315/142/1/012017.
[12] Al-Rashdi, B. Somerfield, C. and Hilal, N. "Heavy metals removal using adsorption and nanofiltration techniques." *Separation & Purification Reviews* 40, no. 3 (2011), 209–259. doi:10.1080/15422119.2011.558165.
[13] Abdel-Fatah, M. A. "Nanofiltration systems and applications in wastewater treatment: review article." *Ain Shams Engineering Journal* 9, no. 4 (2018), 3077–3092. doi:10.1016/j.asej.2018.08.001.
[14] Jumadi, J. Kamari, A. Hargreaves, J. S. and Yusof, N. "A review of nano-based materials used as flocculants for water treatment." *International Journal of Environmental Science and Technology* 17, no. 7 (2020), 3571–3594. doi:10.1007/s13762-020-02723-y.
[15] Yaser, A. Z. Nurmin, B. and Rosalam, S. "Coagulation/flocculation of anaerobically treated palm oil mill effluent (AnPOME): a review." *Developments in Sustainable Chemical and Bioprocess Technology*, 2013, 3–9. doi:10.1007/978-1-4614-6208-8_1.

[16] Badrus, Z. "Potential of natural flocculant in coagulation-flocculation wastewater treatment process." *E3S Web of Conferences* 73 (2018), 05006. doi:10.1051/e3sconf/20187305006.

[17] Hussain, A. Madan, S. and Madan, R. "Removal of heavy metals from wastewater by adsorption." *Heavy Metals—Their Environmental Impacts and Mitigation*, 7, 387-419, (2021). doi:10.5772/intechopen.95841.

[18] Vo, T. S. Hossain, M. M. Jeong, H. M. and Kim, K. "Heavy metal removal applications using adsorptive membranes." *Nano Convergence* 7, no. 1 (2020). doi:10.1186/s40580-020-00245-4.

[19] Gupta, V. K. Moradi, O. Tyagi, I. Agarwal, S. Sadegh, H. Shahryari-Ghoshekandi, R. Makhlouf, A. S. Goodarzi, M. and Garshasbi, A. "Study on the removal of heavy metal ions from industry waste by carbon nanotubes: effect of the surface modification: a review." *Critical Reviews in Environmental Science and Technology* 46, no. 2 (2015), 93–118. doi:10.1080/10643389.2015.1061874.

[20] Luo, J. Fu, K. Yu, D. Hristovski, K. D. Westerhoff, and Crittenden, J. C. "Review of advances in engineering nanomaterial adsorbents for metal removal and recovery from water: synthesis and microstructure impacts." *ACS ES&T Engineering* 1, no. 4 (2021), 623–661. doi:10.1021/acsestengg.0c00174.

[21] Baby, R. Saifullah, B. and Hussein, M. Z. "Carbon nanomaterials for the treatment of heavy metal-contaminated water and environmental remediation." *Nanoscale Research Letters* 14, no. 1, 341-358, (2019). doi:10.1186/s11671-019-3167-8.

[22] Alekseeva, O. V. Bagrovskaya, N. A. and Noskov, A. V. "Sorption of heavy metal ions by fullerene and polystyrene/fullerene film compositions." *Protection of Metals and Physical Chemistry of Surfaces* 52, no. 3 (2016), 443–447. doi:10.1134/s2070205116030035.

[23] Ali, I. Basheer, A. A. Mbianda, X. Y. Burakov, A. Galunin, E. Burakova, I. Mkrtchyan, E. Tkachev, A. and Grachev, V. "Graphene based adsorbents for remediation of noxious pollutants from wastewater." *Environment International* 127 (2019), 160–180. doi:10.1016/j.envint.2019.03.029.

[24] Zhang, C-Z. Chen, B. Bai, Y. and Xie, J. "A new functionalized reduced graphene oxide adsorbent for removing heavy metal ions in water *via* coordination and ion exchange." *Separation Science and Technology* 53, no. 18 (2018), 2896–2905. doi:10.1080/01496395.2018.1497655.

[25] Ahmad, S. Z. N. Salleh, W. N. W. Ismail, A. F. Yusof, Yusop, N. M. Z. M. and Aziz, F. "Adsorptive removal of heavy metal ions using graphene-based nanomaterials: toxicity, roles of functional groups and mechanisms." *Chemosphere* 248 (2020), 126008. doi:10.1016/j.chemosphere.2020.126008.

[26] Alslaibi, T. M. Abustan, I. Ahmad, M. A. and Foul, A. A. "Application of response surface methodology (RSM) for optimization of Cu^{2+}, Cd^{2+}, Ni^{2+}, Pb^{2+}, Fe^{2+}, and Zn^{2+} removal from aqueous solution using microwaved olive stone activated carbon." *Journal of Chemical Technology & Biotechnology* 88, no. 12 (2013), 2141–2151. doi:10.1002/jctb.4073.

[27] Acharya, J. Sahu, J. N. Mohanty, C. R. and Meikap, B. C. "Removal of lead(II) from wastewater by activated carbon developed from Tamarind wood by zinc chloride activation." *Chemical Engineering Journal* 149, no. 1–3 (2009), 249–262. doi:10.1016/j.cej.2008.10.029.

[28] Ismail, A. Harmuni, H. and Mohd, R. R. "Removal of iron and manganese using granular activated carbon and zeolite in artificial barrier of riverbank filtration."

AIP Conference Proceedings, 1835, 020056-020060, (2017). doi:10.1063/1.4983796.

[29] Yang, J. Hou, B. Wang, J. Tian, B. Bi, J. Wang, N. Li, X. and Huang, X. "Nanomaterials for the removal of heavy metals from wastewater." *Nanomaterials* 9, no. 3 (2019), 424. doi:10.3390/nano9030424.

[30] Liang, W. Dai, C. Zhou, X. and Zhang, Y. "Application of zero-valent iron nanoparticles for the removal of aqueous zinc ions under various experimental conditions." *PLoS ONE* 9, no. 1 (2014), e85686. doi:10.1371/journal.pone.0085686.

[31] Tarekegn, M. M. Hiruy, A. M. and Dekebo, A. H. "Nano zero valent iron (nZVI) particles for the removal of heavy metals (Cd^{2+}, Cu^{2+} and Pb^{2+}) from aqueous solutions." *RSC Advances* 11, no. 30 (2021), 18539-18551. doi:10.1039/d1ra01427g.

[32] Danila, V. Vasarevicius, S. and Valskys, V. "Batch removal of Cd(II), Cu(II), Ni(II), and Pb(II) ions using stabilized zero-valent iron nanoparticles." *Energy Procedia* 147 (2018), 214–219. doi:10.1016/j.egypro.2018.07.062.

[33] Lee, S. and Jun, B-H. "Silver nanoparticles: synthesis and application for nanomedicine." *International Journal of Molecular Sciences* 20, no. 4 (2019), 865. doi:10.3390/ijms20040865.

[34] Attatsi, I. K. and Nsiah, F. "Application of silver nanoparticles toward Co(II) and Pb(II) ions contaminant removal in groundwater." *Applied Water Science* 10, no. 6 (2020). doi:10.1007/s13201-020-01240-0.

[35] Al-Qahtani, K. M. "Cadmium removal from aqueous solution by green synthesis zero valent silver nanoparticles with Benjamina leaves extract." *The Egyptian Journal of Aquatic Research* 43, no. 4 (2017), 269–274. doi:10.1016/j.ejar.2017.10.003.

[36] El-Tawil, R. S. El-Wakeel, S. T. Abdel-Ghany, A. E. Abuzeid, H. A. Selim, K. A. and Hashem, A. M. "Silver/quartz nanocomposite as an adsorbent for removal of mercury (II) ions from aqueous solutions." *Heliyon* 5, no. 9 (2019), e02415. doi:10.1016/j.heliyon.2019.e02415.

[37] Murph, S. H. "Nanomaterial-treated filters for the uptake of heavy metals from water sources." (2018). doi:10.2172/1462175.

[38] Qian, H. Pretzer, L. A. Velazquez, J. C. Zhao, Z. and Wong, M. S. "Gold nanoparticles for cleaning contaminated water." *Journal of Chemical Technology & Biotechnology* 88, no. 5 (2013), 735–741. doi:10.1002/jctb.4030.

[39] Naseem, T. and Durrani, T. "The role of some important metal oxide nanoparticles for wastewater and antibacterial applications: a review." *Environmental Chemistry and Ecotoxicology* 3 (2021), 59–75. doi:10.1016/j.enceco.2020.12.001.

[40] Cieschi, M. T. De Francisco, M. Herrero, P. Sánchez-Marcos, J. Cuevas, J. Esteban, E. Lucena, J. J. and Yunta, F. "Synthesis and characterization of nano Fe and Mn (hydr)oxides to be used as natural sorbents and micronutrient fertilizers." *Agronomy* 11, no. 9 (2021), 1876. doi:10.3390/agronomy11091876.

[41] Aragaw, T. A. Bogale, F. M. and Aragaw, B. A. "Iron-based nanoparticles in wastewater treatment: a review on synthesis methods, applications, and removal mechanisms." *Journal of Saudi Chemical Society* 25, no. 8 (2021), 101280. doi:10.1016/j.jscs.2021.101280.

[42] Kumari, P. Alam, M. and Siddiqi, W. A. "Usage of nanoparticles as adsorbents for waste water treatment: an emerging trend." *Sustainable Materials and Technologies* 22 (2019), e00128. doi:10.1016/j.susmat.2019.e00128.

[43] Rajput, S. Pittman, C. U. and Mohan, D. "Magnetic magnetite (Fe_3O_4) nanoparticle synthesis and applications for lead (Pb^{2+}) and chromium (Cr^{6+}) removal from water." *Journal of Colloid and Interface Science* 468 (2016), 334–346. doi:10.1016/j.jcis.2015.12.008.

[44] Jerin, V. M. Remya, R. Thomas, M. and Varkey, J. T. "Investigation on the removal of toxic chromium ion from waste water using Fe_2O_3 nanoparticles." *Materials Today: Proceedings* 9 (2019), 27–31. doi:10.1016/j.matpr.2019.02.032.

[45] Chiavola, A. D'Amato, E. and Boni, M. R. "Comparison of different iron oxide adsorbents for combined arsenic, vanadium and fluoride removal from drinking water." *International Journal of Environmental Science and Technology* 16, no. 10 (2019), 6053–6064. doi:10.1007/s13762-019-02316-4.

[46] Hashemifar, M. G. ShamsKhorramabady, HatamGodini, H. Nilufari, N. Mehrabpour, M. and Davoudi, M. "Preparation of nano iron oxide coated activated sludge granules and its adsorption properties for Cd (II) ions in aqueous solutions." *Research Journal of Environmental and Earth Sciences* 6, no. 5 (2014), 259–265. doi:10.19026/rjees.6.5768.

[47] Chung, K. S. Lee, J. C. Kim, E. J. Lee, K. C. Kim, Y. S. and K. Ooi. "Recovery of lithium from seawater using nano-manganese oxide adsorbents prepared by gel process." *Materials Science Forum* 449–452 (2004), 277–280. doi:10.4028/www.scientific.net/msf.449-452.277.

[48] Lisha, K. P. Maliyekkal, S. M. and Pradeep, T. "Manganese dioxide nanowhiskers: a potential adsorbent for the removal of Hg(II) from water." *Chemical Engineering Journal* 160, no. 2 (2010), 432–439. doi:10.1016/j.cej.2010.03.031.

[49] Al Abdullah, J. Al Lafi, A. G. Al Masri, Yusr Amin, and Tasneem Alnama. "Adsorption of cesium, cobalt, and lead onto a synthetic nano manganese oxide: behavior and mechanism." *Water, Air, & Soil Pollution* 227, no. 7 (2016). doi:10.1007/s11270-016-2938-4.

[50] Ayash, M. A. Elnasr, T. A. and Soliman, M. H. "Removing iron ions contaminants from groundwater using modified nano-hydroxyapatite by nano manganese oxide." *Journal of Water Resource and Protection* 11, no. 06 (2019), 789–809. doi:10.4236/jwarp.2019.116048.

[51] Engates, K. E. and Shipley, H. J. "Adsorption of Pb, Cd, Cu, Zn, and Ni to titanium dioxide nanoparticles: effect of particle size, solid concentration, and exhaustion." *Environmental Science and Pollution Research* 18, no. 3 (2010), 386–395. doi:10.1007/s11356-010-0382-3.

[52] Nabi, D. Aslam, I. and Qazi, I. A. "Evaluation of the adsorption potential of titanium dioxide nanoparticles for arsenic removal." *Journal of Environmental Sciences* 21, no. 3 (2009), 402–408. doi:10.1016/s1001-0742(08)62283-4.

[53] Youssef, A. M. and Malhat, F. M. "Selective removal of heavy metals from drinking water using titanium dioxide nanowire." *Macromolecular Symposia* 337, no. 1 (2014), 96–101. doi:10.1002/masy.201450311.

[54] Cai, Y. Li, C. Wu, D. Wang, W. Tan, F. Wang, X. Wong, P. K. and Qiao, X. "Highly active MgO nanoparticles for simultaneous bacterial inactivation and heavy metal removal from aqueous solution." *Chemical Engineering Journal* 312 (2017), 158–166. doi:10.1016/j.cej.2016.11.134.

[55] Khajeh, M. Laurent, S. and Dastafkan, K. "Nanoadsorbents: classification, preparation, and applications (with emphasis on aqueous media)." *Chemical Reviews* 113, no. 10 (2013), 7728–7768. doi:10.1021/cr400086v.

[56] Zhang, B. Poojary, D. M. Clearfield, A. and Peng, G. "Synthesis, characterization, and amine intercalation behavior of zirconium *n*-(phosphonomethyl)iminodiacetic acid layered compounds." *Chemistry of Materials* 8, no. 6 (1996), 1333–1340. doi:10.1021/cm9506013.

[57] Bruque, S. Aranda, M. A. Losilla, E. R. Olivera-Pastor, P. and Maireles-Torres, P. "Synthesis optimization and crystal structures of layered metal(IV) hydrogen phosphates, .alpha.-M(HPO4)2.cntdot.H2O (M = Ti, Sn, Pb)." *Inorganic Chemistry* 34, no. 4 (1995), 893–899. doi:10.1021/ic00108a021.

[58] Atzrodt, J. Derdau, V. Kerr, W. J. and Reid, M. "Deuterium- and tritium-labelled compounds: applications in the life sciences." *Angewandte Chemie International Edition* 57, no. 7 (2018), 1758–1784. doi:10.1002/anie.201704146.

[59] Yu, J. and Xu, R. "Insight into the construction of open-framework aluminophosphates." *Chemical Society Reviews* 35, no. 7 (2006), 593. doi:10.1039/b505856m.

[60] Qureshi, M. "Comparative study of titanium(IV)-based exchangers in aqueous and mixed solvent systems." *Talanta* 20, no. 7 (1973), 609–620. doi:10.1016/0039-9140(73)80111-0.

[61] Tranter, T. J. Herbst, R. S. Todd, T. A. Olson, A. L. and Eldredge, H. B. "Evaluation of ammonium molybdophosphate-polyacrylonitrile (AMP-PAN) as a cesium selective sorbent for the removal of 137Cs from acidic nuclear waste solutions." *Advances in Environmental Research* 6, no. 2 (2002), 107–121. doi:10.1016/s1093-0191(00)00073-3.

[62] Suss S. M. and Pfrepper, G. "Investigations of the sorption of cesium from acid solutions by various inorganic sorbents." *Radiochimica Acta*, 29, no. 1 (1981), 33–40. doi:10.1524/ract.1981.29.1.33.

[63] Chen, Z. Wu, Y. Wei, Y. and Mimura, H. "Preparation of silica-based titanate adsorbents and application for strontium removal from radioactive contaminated wastewater." *Journal of Radioanalytical and Nuclear Chemistry* 307, no. 2 (2015), 931–940. doi:10.1007/s10967-015-4470-1.

[64] Zhang, M. Jin, Z. Zhang, J. Guo, X. Yang, J. Li, W. Wang, X. and Zhang, Z. "Effect of annealing temperature on morphology, structure and photocatalytic behavior of nanotubed H2Ti2O4(OH)2." *Journal of Molecular Catalysis A: Chemical* 217, no. 1–2 (2004), 203–210. doi:10.1016/j.molcata.2004.03.032.

[65] Du, X. Xu, Y. Ma, H. Wang, J. and Li, X. "Synthesis and characterization of bismuth titanate by an aqueous sol?gel method." *Journal of the American Ceramic Society* 90, no. 5 (2007), 1382–1385. doi:10.1111/j.1551-2916.2007.01548.x.

[66] Wen, P. Ishikawa, Y. Itoh, H. and Feng, Q. "Topotactic transformation reaction from layered titanate nanosheets into anatase nanocrystals." *The Journal of Physical Chemistry C* 113, no. 47 (2009), 20275–20280. doi:10.1021/jp908181e.

[67] Zou, Y. Wang, P. Yao, W. Wang, X. Liu, Y. Yang, D. Wang, L. et al. "Synergistic immobilization of UO_2^{2+} by novel graphitic carbon nitride @ layered double hydroxide nanocomposites from wastewater." *Chemical Engineering Journal* 330 (2017), 573–584. doi:10.1016/j.cej.2017.07.135.

[68] Yao, W. Wang, J. Wang, P. Wang, X. Yu, S. Zou, Y. Hou, J. Hayat, T. Alsaedi, A. and Wang, X. "Synergistic coagulation of GO and secondary adsorption of heavy metal ions on Ca/Al layered double hydroxides." *Environmental Pollution* 229 (2017), 827–836. doi:10.1016/j.envpol.2017.06.084.

[69] Yao, W. Wang, X. Liang, Y. Yu, S. Gu, P. Sun, Y. Xu, C. et al. "Synthesis of novel flower-like layered double oxides/carbon dots nanocomposites for U(VI) and

241Am(III) efficient removal: Batch and EXAFS studies." *Chemical Engineering Journal* 332 (2018), 775–786. doi:10.1016/j.cej.2017.09.011.

[70] Wang, X. Ning, S. Liu, R. and Wei, Y. "Stability of isoHex-BTP/SiO$_2$-P adsorbent against acidic hydrolysis and γ-irradiation." *Science China Chemistry* 57, no. 11 (2014), 1464–1469. doi:10.1007/s11426-014-5200-1.

[71] Ning, S. Zou, Q. Wang, X. Liu, R. Wei, Y. Zhao, Y. and Ding, Y. "Evaluation study on silica/polymer-based CA-BTP adsorbent for the separation of minor actinides from simulated high-level liquid wastes." *Journal of Radioanalytical and Nuclear Chemistry* 307, no. 2 (2015), 993–999. doi:10.1007/s10967-015-4248-5.

[72] Chen, Z. Wu, Y. and Wei, Y. "Adsorption characteristics and radiation stability of a silica-based DtBuCH18C6 adsorbent for Sr(II) separation in HNO3 medium." *Journal of Radioanalytical and Nuclear Chemistry* 299, no. 1 (2013), 485–491. doi:10.1007/s10967-013-2750-1.

[73] Chen, Z. Wu, Y. and Wei, Y. "The effect of temperatures and γ-ray irradiation on silica-based calix[4]arene-R14 adsorbent modified with surfactants for the adsorption of cesium from nuclear waste solution." *Radiation Physics and Chemistry* 103 (2014), 222–226. doi:10.1016/j.radphyschem.2014.06.004.

[74] de Oliveira, A. D. and Beatrice, C. A. G. "Polymer nanocomposites with different types of nanofiller." *Nanocomposites—Recent Evolutions* (2019). doi:10.5772/intechopen.81329.

[75] Kumar, R. Rauwel, P. and Rauwel, E. "Nanoadsorbants for the removal of heavy metals from contaminated water: current scenario and future directions." *Processes* 9, no. 8 (2021), 1379. doi:10.3390/pr9081379.

[76] Borji, H. Ayoub, G. M. Bilbeisi, R. Nassar, N. and Malaeb, L. "How effective are nanomaterials for the removal of heavy metals from water and wastewater?" *Water, Air, & Soil Pollution* 231, no. 7 (2020). doi:10.1007/s11270-020-04681-0.

[77] Wahab M. A. Othaman, R. and Hilal, N. "Potential use of nanofiltration membranes in treatment of industrial wastewater from Ni-P electroless plating." *Desalination* 168 (2004), 241–252. doi:10.1016/j.desal.2004.07.004.

[78] Chen, J. Qiu, X. Fang, Z. Yang, M. Pokeung, T. Gu, F. Heng, W. and Lan, B. "Removal mechanism of antibiotic metronidazole from aquatic solutions by using nanoscale zero-valent iron particles." *Chemical Engineering Journal* 181–182 (2012), 113–119. doi:10.1016/j.cej.2011.11.037.

[79] Liu, T. Wang, Z-L. Yan, X. and Zhang, B. "Removal of mercury (II) and chromium (VI) from wastewater using a new and effective composite: pumice-supported nanoscale zero-valent iron." *Chemical Engineering Journal* 245 (2014), 34–40. doi:10.1016/j.cej.2014.02.011.

[80] Zhang, Y. Zhang, Y. Akakuru, O. U. Xu, X. and Wu, A. "Research progress and mechanism of nanomaterials-mediated in-situ remediation of cadmium-contaminated soil: a critical review." *Journal of Environmental Sciences* 104 (2021), 351–364. doi:10.1016/j.jes.2020.12.021.

[81] Nassar, N. N. "Rapid removal and recovery of Pb(II) from wastewater by magnetic nanoadsorbents." *Journal of Hazardous Materials* 184, no. 1–3 (2010), 538–546. doi:10.1016/j.jhazmat.2010.08.069.

[82] Ağtaş, M. Ormancı-Acar, T. Keskin, B. Türken, T. and Koyuncu, İ. "Nanofiltration membranes for salt and dye filtration: effect of membrane properties on performances." *Water Science and Technology* 83, no. 9 (2021), 2146–2159. doi:10.2166/wst.2021.125.

[83] Zand, A. D. Tabrizi, A. M. and Heir, A. V. "Application of titanium dioxide nanoparticles to promote phytoremediation of Cd-polluted soil: contribution of PGPR inoculation." *Bioremediation Journal* 24, no. 2–3 (2020), 171–189. doi:10.1080/10889868.2020.1799929.

[84] Fu, R. Yang, Y. Xu, Z. Zhang, X. Guo, X. and Bi, D. "The removal of chromium (VI) and lead (II) from groundwater using sepiolite-supported nanoscale zero-valent iron (S-NZVI)." *Chemosphere* 138 (2015), 726–734. doi:10.1016/j.chemosphere.2015.07.051

[85] Senthil K. Sivaranjanee, P. R. Sundar, R. P. and Saravanan, A. "Carbon sphere: synthesis, characterization and elimination of toxic Cr(VI) ions from aquatic system." *Journal of Industrial and Engineering Chemistry* 60 (2018), 307–320. doi:10.1016/j.jiec.2017.11.017.

[86] Sadegh, H. Ali, G. A. Gupta, V. K. Makhlouf, A. S. Shahryari-ghoshekandi, R. Nadagouda, M. N. Sillanpää, M. and Megiel, E. "The role of nanomaterials as effective adsorbents and their applications in wastewater treatment." *Journal of Nanostructure in Chemistry* 7, no. 1 (2017), 1–14. doi:10.1007/s40097-017-0219-4.

[87] Malik, R. Bhaskaran, M. and Lata, S. "Heavy metal removal from wastewater using adsorbents." *Environmental Chemistry for a Sustainable World* (2020), (9) 441–469. doi:10.1007/978-3-030-52421-0_13.

7 Photocatalytic Degradation of Organic Pollutants by Using Efficient Nanomaterials

Vijayalakshmi Gosu,[1] Gayatri Rajpur,[1] Uttam Singh,[1] Meena Nemiwal,[2] S. Suresh,[3] and Verraboina Subbaramaiah[1]*

[1]Department of Chemical Engineering, Malaviya National Institute of Technology Jaipur, Jaipur 302017, India
[2]Department of Chemistry, Malaviya National Institute of Technology Jaipur, Jaipur 302017, India
[3]Department of Chemical Engineering, Maulana Azad National Institute of Technology, Bhopal 462003, India
*Corresponding author: Email: vlakshmi.chem@mnit.ac.in (VG), vsr.chem@mnit.ac.in (VS)

CONTENTS

7.1 Introduction 126
7.2 Degradation Mechanism of Organic Pollutants with Photocatalytic Nanomaterials 128
7.3 Synthesis Methods for Photocatalytic Nanomaterials 131
    7.3.1 Synthesis of Photo Nanocatalyst Using Ball Milling 131
    7.3.2 Synthesis of Photo Nanocatalyst Using Sol-Gel 131
    7.3.3 Synthesis of Photo Nanocatalysts Using an Impregnation Method 132
    7.3.4 Synthesis of Photo Nanocatalysts Using Hydrothermal and Solvothermal Processes 133
    7.3.5 Synthesis of Photo Nanocatalysts Using a Chemical Vapor Deposition Method (CVD) 134
7.4 Photo Nanocatalyst Materials 137
    7.4.1 ZnO Based Photo Nanocatalyst 137
    7.4.2 TiO_2 Based Photo Nanocatalyst 138
    7.4.3 Carbonaceous Based Photo Nanocatalyst 139

DOI: 10.1201/9781003252931-7

                7.4.4   Metal-Organic Framework Based Photo Nanocatalyst 139
                7.4.5   Doping and Co-doping of Metals in Semiconductor
                        Frameworks 140
                7.4.6   Iron-Based Photo Nanocatalyst 140
        7.5   Conclusion 141
        References 141

7.1 INTRODUCTION

Water is essential for life on Earth and no life without water. In a nutshell, water is a very valued gift of nature. Expanding rapid industrialization and urbanization create a huge demand for freshwater requirements, consequently outputting a large quantity of Wastewater into the environment. Wastewater derived from domestic, industrial, and agricultural activities typically consists of organic-rich waste water. Organic pollutants typically comprise phenols, biphenyls, plasticizers, pyridines, antibiotics, hydrocarbon solvents, pesticides, herbicides, dyes, pharmaceuticals, and so on [1].

Organic pollutants are of great concern due to their toxicity, persistence, and concentration, and have major environmental, public health, and economic consequences. These organic contaminants have low octanol/water partition coefficients (POW) and are commonly found in manufacturing industrial wastewater. As a result, bioavailability rises, potentially posing a greater environmental risk [2]. Numerous classical wastewater treatment technologies have been widely explored to treat organic pollutant-containing wastewater, including coagulation, flocculation, adsorption, reverse osmosis, incineration, and activated sludge [3,4]. Hazardous refractory organic wastewater is incompatible with traditional biological treatment methods.

These methods require a long residence time to treat refractory pollutants [4]. Incineration is considered a powerful treatment. However, it requires a lot of energy to destroy the organic contaminants.

Harmful secondary pollutants such as dioxins and furans are released into the environment because of this process [5]. Despite recent advances, several of these treatments still have significant limitations, including high operating costs, low efficiency, and the additional burden of secondary sludge treatment [6]. Therefore, exploring effective treatment technologies is a significant research challenge for organic wastewater treatment.

Advanced oxidation processes (AOPs) are an innovative method for degrading hazardous organic wastes. AOPs usually produce potent oxidizing radicals that break down organic molecules non-selectively into harmless compounds. The photo-catalytic process has gained a lot of interest among the many AOPs since it is considered a sustainable technology because it has a low operating cost, requires no additional oxidants, and operates under ambient conditions. Photocatalysts are widely used in many reactions to speed up reaction rates, and not just for

Photocatalytic Degradation of Organic Pollutants

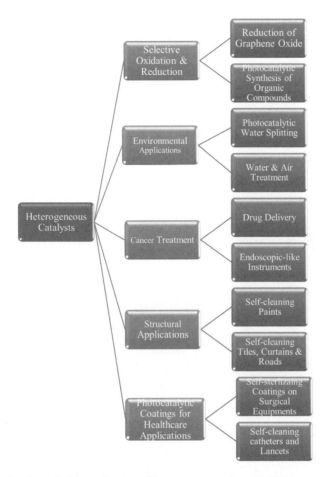

FIGURE 7.1 Conceivable application of heterogeneous photocatalysis.

environmental remediation (Figure 7.1) [7]. However, there is still a long way to go in order to develop a catalyst with high selectivity and reactivity. As a result, there is a pressing need to investigate novel materials and methods for producing efficient photocatalysts.

Nanotechnology is a fast-growing field of study and development that is broad and interdisciplinary. Nanomaterials have recently aided in developing and implementing novel and cost-effective materials to degrade organic wastes (EPA, 2005). Various nanomaterials with characteristic photocatalytic activity, such as ZnO, Fe_2O_3, Fe_3O_4, WO_3, Bi_2WO_6, Ag_3PO_4, CdS, In_2S_3, MoS_2, $ZnSeC_3N_4$, AgCl, BiOCl, and others (TiO_2, WO_3, and so forth) have been examined to aid the degradation of organic-containing wastewater by numerous researchers.

7.2 DEGRADATION MECHANISM OF ORGANIC POLLUTANTS WITH PHOTOCATALYTIC NANOMATERIALS

The photocatalytic approach is expected to provide a long-term solution for the mineralization of organic wastewater. Figure 7.2 depicts the fundamental properties of a promising photocatalyst. The in-situ generation of oxidation radicals at the conduction and valance bands of the surface of the catalyst has improved the mineralization of organic waste. A series of oxidative and reductive reactions on the surfaces of nanomaterials started the mineralization process. Light intensity and the bandgap of the catalyst also influenced the rate of the photocatalytic process [8]. Photocatalysts can be homogeneous or heterogeneous. In homogeneous photocatalysis, metal complexes such as iron, copper, chromium, and others are used as catalysts in presence light to form the hydroxyl radical (•OH), which degrades organic contaminants non-selectively [1]. Separation and recycling of homogeneous catalysts are extremely difficult.

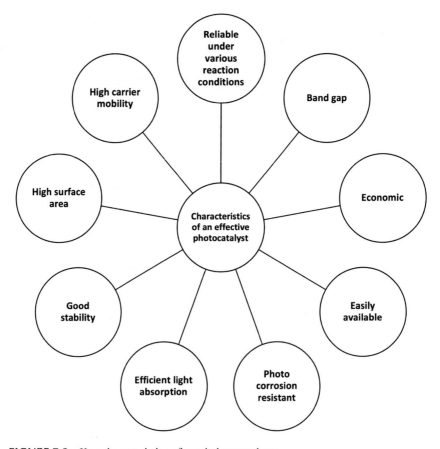

FIGURE 7.2 Key characteristics of good photocatalysts.

TABLE 7.1
Band Gap Energy of Photocatalyst

Photocatalyst	Bandgap Energy (eV)	Photocatalyst	Bandgap Energy (eV)
TiO_2	3.2	Diamond	5.4
GaP	2.2	Cu_2O	2.172
ZrO_2	2.2	ZnS	3.60
Si	1.1	PbS	0.286
CdS	2.5	PbSe	0.165
ZnO	3.2	CdSe	1.7
Fe_2O_3	2.2	Si	1.170
WO_3	2.5	Ge	0.744
SnO_3	3.5		

Their potential discharge into the environment may pose significant threats to ecological systems and human health. An efficient way to overcome the this technical bottleneck is to use heterogeneous photocatalysts, gaining popularity because of their high oxidation power, stability, biological and chemical inertness, low cost, and non-toxicity. Various photocatalysts, including TiO_2, ZnO, Wo_3, CdS, and V_2O_5, have been widely used for the photocatalytic destruction of organic pollutants in wastewater. Nanoscale alteration of catalytic materials has boosted the efficiency of heterogeneous catalysts. These nanoscale materials significantly reduce the bandgap and electron recombination rate, enhancing degradation efficiency. Semiconductor nanomaterials are excited with light in the highest occupied energy band, known as the valence band, and the lowest unoccupied energy band, known as the conduction band. The bandgap is the difference in energy levels between the conduction and valence bands. Table 7.1 describes the bandgap of several photocatalysts. The major steps involved in the photocatalysis process to degrade organic pollutants are shown in Figure 7.3 [9].

Step-1: Light energy (photons) is irradiated onto the surface of a semiconductor. If the incident photon energy is equal to or larger than the semiconductor's bandgap, electrons in the valance band are photo-excited and transferred to the conduction band. As a result, photogenerated electrons (e-) are found in the conduction band, while created holes (h^+) are found in the valance band. The generation of the electron-hole (e^-) - (h^+) pair is shown in Eq-1 on the surface of the photocatalyst. Step-2: The holes formed in the valance band oxidize the donor molecule, which reacts with the available water molecule to produce the powerful hydroxyl radicals illustrated in Eq-6, responsible for the non-selective destruction of the organic pollutant. Step-3: Electrons in the conduction band interact with oxygen, generating superoxide ions and causing the redox processes depicted in Figure 7.3. On the surface of the semiconductor, the electron-hole pair undergoes a sequence of oxidative and reductive processes. However, there are several limitations to the

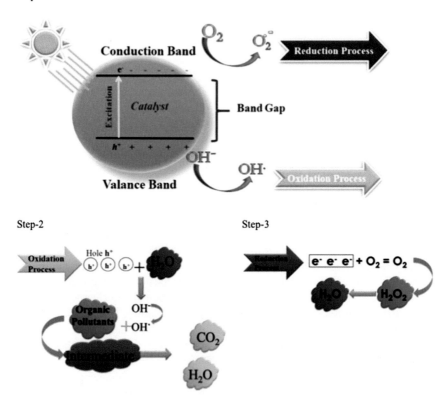

FIGURE 7.3 Mechanism of photocatalytic degradation of organic pollutants.

photocatalytic process, such as interfacial charge transfer, charge separation, and charge carrier recombination. As a result, innovative alternative nanomaterials and surface modification on semiconductors are greatly needed to overcome restrictions. Many authors investigated various nanostructured photocatalysts for organic wastewater degradation.

$$TiO_2 + hv \rightarrow TiO_2(e_{CB}^- + h_{VB}^+) \tag{1}$$

$$TiO_2(h_{VB}^+) + H_2O \rightarrow TiO_2 + H^+ + OH^- \tag{2}$$

$$TiO_2(h_{VB}^+) + OH^- \rightarrow TiO_2 + \bullet OH \tag{3}$$

$$TiO_2(e_{CB}^-) + O_2 + H^+ \rightarrow TiO_2 + HO_2^\bullet \rightarrow O_2^{\bullet-} + H^+ \tag{4}$$

$$TiO_2(e_{CB}^-) + HO_2^{\bullet} + H^+ \rightarrow H_2O_2 \quad (5)$$

$$TiO_2(e_{CB}^-) + HO_2^{\bullet} \rightarrow TiO_2 + OH^- + \bullet OH \quad (6)$$

7.3 SYNTHESIS METHODS FOR PHOTOCATALYTIC NANOMATERIALS

7.3.1 Synthesis of Photo Nanocatalyst Using Ball Milling

Ball milling is one of the top-down methodologies to produce nanomaterials/photocatalysts. The ball milling method is typically used to grind and blend materials to reduce the size of bulk solid particles to meso and nanoscales. A ball mill is usually made of a hollow cylindrical shell around its axis. The axis of the shell can be horizontal or at an inclination to the horizontal. The crushed media typically comprises of several sizes of balls (steel (chrome steel), stainless steel, ceramic, or rubber). Sapkota and Mishra [10] employed ball milling to generate the nanomaterial p-CuO/n-ZnO, which was then used to degrade MB dye via a photo catalytic process.

7.3.2 Synthesis of Photo Nanocatalyst Using Sol-Gel

Sol-gel synthesis is one of the most frequently used techniques for synthesizing photo-catalysts. It has recently acquired popularity because of its easy preparation, low cost, and mild reaction conditions. This method is used to prepare nanometal-oxides. Figure 7.4 illustrates a schematic illustration of the sol-gel process for nanomaterial synthesis. In this technique, metals, or metal alkoxides are dissolved in a suitable solvent with continuous stirring, followed by adding a precipitating agent drop by drop while stirring vigorously, gel is formed as a result of the precipitation. A result, precipitate gel forms. To eliminate solvents, present in the precipitate, the precipitate (gel) should be dried at a temperature slightly above the solvent's boiling point. The collected NPs are thermally treated at the desired temperature to produce photocatalytic nanomaterials. Behnajady et al. [11] used the sol-gel method to synthesize TiO_2 NPs and then tested the organic pollutant's photocatalytic activity (C.I. Acid Red 27). Ahmed et al. [12] employed the sol-gel process to prepare Fe_2O_3/TiO_2 NPs for the photocatalytic degradation of methylene blue dye. The physical and chemical properties of the sol-gel process product are strongly influenced by the process parameters such as precursor type, the rate of hydrolysis, the time of aging, the pH, and the molar ratio of precursor and water, and so forth. Photocatalytic nanomaterial prepared by sol-gel methods and their application in the degradation of organic pollutants is described in Table 7.2.

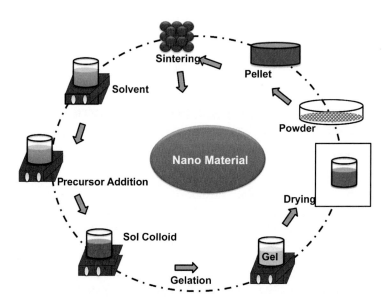

FIGURE 7.4 Schematic representation of Photo Nanocatalysts synthesis using the sol-gel process.

7.3.3 Synthesis of Photo Nanocatalysts Using an Impregnation Method

Impregnation is the most promising and extensively used method to prepare photocatalytic nanomaterials. This technique has been used to generate an active nanometal-incorporated porous catalyst. The required amount of active metal precursor is dissolved in the solvent, followed by the addition of support material to the precursor solution. The resulting mixture was dried and calcined to produce a metal oxide impregnated porous photocatalyst. Wet impregnation is frequently performed using an excess solvent to dissolve the precursor solution, which is typically much greater than the pore volume of the support material. The excess quantity of precursor solution employed in the wet impregnation procedure is recycled to limit waste generation in industrial-scale production. The schematic illustration of photo nanocatalyst synthesis using the impregnation process is displayed in Figure 7.5.

However, a fixed amount of solvent equal to the support's pore volume has been deployed during the dry or incipient wetness impregnation procedure. Adopting the incipient wet impregnation technique avoids the second filtration step, and it minimizes extra precursor formation. The impregnation technique has been used to produce nano photocatalytic materials. The fundamental disadvantage of impregnation is the lack of control over the size of the metal precursor. Table 7.2 represents photocatalytic nanomaterial prepared by

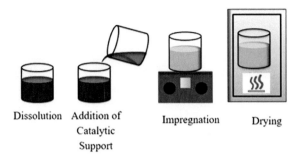

FIGURE 7.5 Schematic illustration of the photo nanocatalyst synthesis using the impregnation process.

impregnation and their application in the degradation of organic pollutants. Mizukoshi et al. [13] immobilized noble metal NPs on the surface of TiO_2 through the impregnation process. Behnajady and Eskandarloo [14] used co-impregnated silver and copper NPs on the surface of TiO_2-P25 NPs for photocatalytic degradation of organic contaminants. The photocatalyts prepared from impregnation process for the degradation of organic pollutants are depicted in Table 7.2.

7.3.4 Synthesis of Photo Nanocatalysts Using Hydrothermal and Solvothermal Processes

The hydrothermal method is one of the most widely used for synthesizing nanomaterials. Figure 7.6 illustrates the hydrothermal process for synthesizing Photo Nanocatalyst. The active metal precursor is dissolved into the solvent, and then the precipitating agent is added drop by drop. The resultant residue was hydrothermally treated in an autoclave at a high temperature (130 - 250°C) and high vapor pressure (0.3 - 4 MPa) for 1 - 12 hours. After hydrothermal treatment, the resultant powder was cooled and washed with the washing media. The resulting nanostructured materials were dried for 24 hours at 80°C.

Although the solvothermal and hydrothermal methods are similar, the main difference is that the hydrothermal approach uses an aqueous solution throughout the crystallization process and the solvothermal method uses a non-aqueous solution. Zulfiqar et al. [15] investigated the hydrothermal synthesis method to synthesize multiwalled TiO_2 nanotubes to treat organic pollutants (Orange II). Munoz-Fernandez et al. [16] used the solvothermal method to synthesize Ag/ZnO and Pt/ZnO nanocomposites and then used them for dye degradation. Photocatalytic nanomaterial was prepared by hydrothermal methods. The hydrothermal and solvothermal processes and their degradation of organic pollutants are presented in Table 7.2.

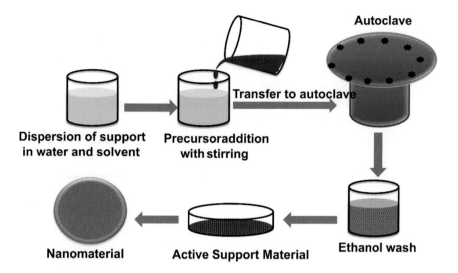

FIGURE 7.6 Schematic representation to synthesize the photo nanocatalysts by hydrothermal and solvothermal process.

7.3.5 Synthesis of Photo Nanocatalysts Using a Chemical Vapor Deposition Method (CVD)

For preparing nanomaterial, CVD is one of the well-known synthesis techniques, thin film deposition is one of its most popular applications, that is used for the preparation of nanomaterial. CVD is a vacuum deposition technology for producing high-quality and high-performance nanomaterials. The nanomaterials prepared using CVD are utilized in several industries for high-temperature protection, erosion protection, and a combination of the two. Thin films are frequently produced with this method in the semiconductor sector. The film is deposited on the substrate's surface via a gas-phase or vapor-phase precursor with or without chemical reaction. A substrate, typically a semiconductor, is coated with a thin film of various precursors such as solid, liquid and gas. The precursor may be made up of one or more components, but it is always employed in a gaseous state. Figure 7.7 displays the outline of the preparation of nanomaterial by a chemical vapor deposition process. The reaction is always carried out in an inert atmosphere at high temperature and pressure. The thin film deposited on the surface of the substrate either reacts or decomposes with the aid of precursors. There are various CVD methods for the deposition depending upon the type of the precursor (liquid/gas phase) and substrate.

This technique provides better control of over growth and structural and architectural parameters. The photo-catalytic nanomaterial prepared by the chemical vapor deposition method and the degradation of organic pollutants is depicted in Table 7.2. Sun et al. [17] prepared TiO_2 films by chemical vapor deposition at atmospheric pressure. Thirumal et al. [18] employed the facile single-step synthesis of MXene@CNTs hybrid nano composite by a CVD method to degrade hazardous pollutants.

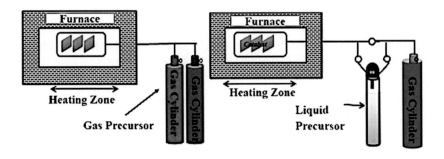
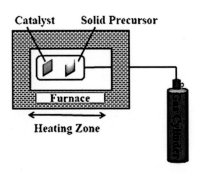

FIGURE 7.7 Schematic outline of preparing nanomaterial by chemical vapor deposition process.

TABLE 7.2
Photo-catalytic Nanomaterial Prepared by Various Synthesis Methods and Their Application in the Degradation of Organic Pollutants

Photo-catalytic Nanomaterial	Organic Pollutant	Light Source	% Removal	Reference
Sol-gel				
Mg^{2+} or Ba^{2+} metal ion doped Nano TiO_2	4-Chlorophenol	Ultraviolet	-	[19]
Zn doped TiO_2	4-Nitrophenol	UV-Visible light	91.6	[20]
Co-TiO_2	*p*-Nitrophenol	Visible light	50	[21]
$ZnTiO_3$	methyl orange	Ultraviolet light	70	[22]
Nickel oxide (NiO)	Phenol	Ultraviolet laser	97	[23]
Ag–CdS@Pr-TiO_2	Methyl Orange	Visible light	-	[24]

(*continued*)

TABLE 7.2 (Continued)
Photo-catalytic Nanomaterial Prepared by Various Synthesis Methods and Their Application in the Degradation of Organic Pollutants

Photo-catalytic Nanomaterial	Organic Pollutant	Light Source	% Removal	Reference
Sm doped TiO_2	Methyl Orange	Ultraviolet light	96	[24]
ZnO	Methylene blue	Ultraviolet light	92	[25]
2%Fe–ZnO	Methylene blue	Ultraviolet light	92	[26]
$NiMoO_4$	Methylene Blue	UV-Visible light	-	[27]
Ba doped $BiFeO_3$	Benzene	Visible light	97%	[28]
Impregnation				
Cu_2O-CeO_2-ZnO	Tolytriazol	Visible light	-	[29]
Cu-Ni/TiO_2	DIPA	Visible light	86.82	[30]
Cu_2O-CeO_2-ZnO	Tolytriazol	Visible light	-	[29]
Fe_2O_3/TiO_2	Remazol Dye	UV light	90.57	[31]
Sodium Alignate doped with TiO_2	Methyl Orange		97.9	[32]
Sodium Alignate doped with a mixture of TiO_2 and SiO_2	Methyl Orange	Sun Light	73.5	[32]
PVA/TiO_2 hybrid film	Methyl Orange	UV light	-	[33]
ZrO_2/MoS_2 (MZr)	Methyl Orange	UV light	95.1	[34]
Hydrothermal and Solvothermal				
CdS/CuS	Methyl Orange		93	[35]
TiO_2@β-GDY	Methyl Blue	-	-	[36]
GDY-ZnO	RhB and MB	-	MB = 80 RhB = 55	[37]
Au@$CaTiO_3$	Rhodamine B	UV- Visible light	99.6	[38]
BiOI/MIL-125(Ti)	Tetracycline	Visible light	80	[39]
NH_2-MIL-101(Fe)	Toluene	Visible light	79.4	[40]
CdS/MIL-53(Fe)	Rhodamine B	Visible light	92.5	[35]
Mn-doped $SrTiO_3$	Tetracycline	Visible light	66.7	[41]
Graphene oxide-TiO_2	Tetrabromobisphenol	Ultraviolet	95.9	[42]
TiO_2/MoS_2	Methyl orange	Visible light	97	[43]
MgO@ CNT	Sulfadiazine	Visible light	96	[44]
CoO@GR	TBBPA	Visible light	69% in 120 min	[45]

TABLE 7.2 (Continued)
Photo-catalytic Nanomaterial Prepared by Various Synthesis Methods and Their Application in the Degradation of Organic Pollutants

Photo-catalytic Nanomaterial	Organic Pollutant	Light Source	% Removal	Reference
Titanate nanotube/ single-wall carbon nanotube	4-Chlorophenol	Ultraviolet and solar irradiation	100% - 120 min 100% - 240 min	[46]
Chemical vapor deposition method (CVD)				
ZnO nanostructures on Si substrates	Rhodamine B	Ultraviolet	-	[47]
$Fe_7S_8/Fe_3O_4@Fe$	municipal sewage	Ultraviolet and solar irradiation	-	[48]
TNT-PAN	reactive black 5	visible	56.4%	[49]
Ti-coating MgAl hydrotalcite	methylene blue	Ultraviolet		[50]
Sulfur-doped TiO_2 films on borosilicate substrate	methyl orange	visible	52%	[51]

7.4 PHOTO NANOCATALYST MATERIALS

This chapter examines photocatalyst materials such as TiO_2, CdS, Fe_2O_3, ZnO, metal-organic frameworks, and metal and nonmetal doping have been investigated to break organic molecules into harmless substances. This review report may be a good information resource for new researchers interested in photocatalytic degradation of organic pollutant.

7.4.1 ZnO Based Photo Nanocatalyst

ZnO has traditionally been utilized in catalytic reactions, including photocatalytic processes. ZnO has conventional semiconductor properties, and many researchers synthesized the ZnO NPs for photocatalytic destruction of organic wastes. ZnO is a semiconductor of the n-type II-IV family. The features of ZnO NPs as follows 3.37 eV bandgap, a high excitation binding energy of 60 meV, and good visible transmission. ZnO has various advantages, such as low cost, environmental safety, high activity rates, and many active sites with high surface reactivity. However, the ZnO catalyst is only active in UV radiation because of its high bandgap of 3.37 eV. Many researchers have circumvented these obstacles by structurally altering ZnO NPs and doping the ZnO structure with metals and nonmetal species.

To maximize photocatalysis efficiency, the electron-hole pair recombination rate must be kept as low as possible. Recently, scientists attempted to explore an alternative route to minimize the electron-hole pair recombination rate by introducing doping materials such as metals and nonmetals into the semiconductor framework. The dopants trap the electrons, reducing the electron-hole pair recombination rate and increasing the photocatalytic system's activity. The sufficiently large surface area photocatalyst provides adequate catalytic activity for the photocatalytic reaction. Various dopants such as transition metals (Fe, Cu), alkaline earth metals (Mg, Sr), noble metals (Au, Ag, Pd), rare earth metals (Eu, Nd, Er), and nonmetals (S, C, N) have been incorporated in a ZnO framework to improve the photocatalytic activity of the ZnO photocatalyst. The nanostructured ZnO catalyst, with dopant, performed excellent photocatalytic activity to degrade organic wastewater.

7.4.2 TiO_2 Based Photo Nanocatalyst

Titanium dioxide has been intensively utilised for mineralizing organic or antibiotic compound-containing wastewater and discoloration of dye effluent. TiO_2 has a bandgap of 3.4 eV, which is equivalent to the wavelength of UV light, and it has excellent physical and chemical properties. Since a significant portion of the solar spectrum comprises visible light (44%), only 5% of the solar spectrum comprises UV light. Nearly 40% of sunlight is composed of visible photons, with the remaining 4 - 5% composed of UV photons. Semiconductors frequently have a wide bandgap and absorb light in the ultraviolet (400 nm) spectrum. As a result, the usefulness of bare TiO_2 has been limited [52]. Mere TiO_2 utilization hampered its scaleup possibilities because of the fast recombination of e-/h+ pairs. The main strategies for developing visible-light-responsive TiO_2-based photocatalysts are element doping and heterostructure fabrication.

Many methods have been developed to increase the photocatalytic performance of semiconductors in the visible light range. Although large bandgap materials are incapable of absorbing visible light, tiny bandgap semiconductor materials can sensitize them, allowing the newly formed composite system to absorb visible light because of the strong coupling effect. Several scientific strategies have been developed to work around these limits. Integrating narrow bandgap materials into the semiconductor structure allows for absorption in the visible or infrared ranges [53]. Nanorods, nanotubes, nanosheets, and nanofibers are some of the nanostructured TiO_2 materials that have been employed in photocatalytic degradation. Hydrothermal decomposition, chemical vapor deposition, sol-gel precipitation, and solvothermal procedures are the most extensively used methods for nano TiO_2 synthesis [54]. Introducing iron species into the nano TiO_2 framework caused a shift in the energy levels of the valance and conduction bands, resulting in intermediate energy levels that allowed the Fe-TiO_2 nanocomposite to excite in the visible range [54]. Several scientific strategies have been developed to work around these limits. Now a days, many authors are developing novel photoactive materials with improved catalytic properties.

7.4.3 Carbonaceous Based Photo Nanocatalyst

To boost the photocatalytic ability of semiconductors, simple semiconductors have been altered with carbon-based materials. Several carbon-based compounds such as graphene, graphite, carbon nanotubes, and nanodiamonds have been employed to alter semiconductor materials. The resulting photocatalyst has a large surface area, appropriate surface chemistry, and great chemical stability. Carbon-based semiconductors demonstrate potential catalytic activity for organic wastewater degradation. Carbonaceous nanomaterials such as GO, rGO, graphene quantum dots, CNTs, and carbon quantum dots are promising candidates for heterostructure photocatalyst fabrication. These materials have a variety of advantages, including flexible bandgap energy, fast electron transfer capacity, minimal toxicity, and an inexpensive cost.

Semiconducting carbonaceous nanomaterials can operate as electron sinks, trapping electrons from the semiconductor and reducing the rate of electron-hole pair recombination. In addition, the heterojunction carbonaceous photocatalyst enhances visible-light harvesting. Several investigations into the conjugation of semiconductor materials with new carbonaceous nanomaterials have been conducted throughout the previous decade. Wu et al. [55] explored the application of the heterojunction photocatalyst, $TiO_2/g-C_3N_4$, to degrade aqueous formaldehyde. They found that the heterojunction photocatalyst retains redox characteristics and increases photocatalytic performance by lowering electron-hole pair recombination. Carbonaceous materials have been used as heterojunction photocatalysts for the visible-light-driven photocatalytic degradation of water contaminants.

7.4.4 Metal-Organic Framework Based Photo Nanocatalyst

Traditional photocatalytic materials such as TiO_2, ZnO, and SnO_2 and their derivatives have shown excellent photocatalytic activity in degrading organic pollutants. Recently, scientists have been exploring metal-organic frameworks (MOFs) for their feasibility for the photocatalytic degradation of organic waste water. Zhang et al. [56] synthesized MOFs (hexagonal NH_2-MIL-101 (Fe) microspindles) for the photodegradation of toluene vapor in the presence of visible light. The addition of amino functionality enabled broad and efficient adsorption in the visible light band.

In the mid-1990s, metal-organic frameworks (MOFs) were constructed (MOF-5) and have been used for the photodegradation of phenol-bearing wastewater. MOFs are typically porous crystalline materials constructed with metal ions and organic ligands. These materials possess high porosity, large surface area, customizable function, and tailorable structure. Recently, many authors have explored the perspective of MOFs as photocatalysts for the degradation of organic waste water [57]. Subsequently, many authors have made great efforts to explore the potential of MOFs as a new class of photocatalysts.

7.4.5 DOPING AND CO-DOPING OF METALS IN SEMICONDUCTOR FRAMEWORKS

Recently, scientists have focused their attention on doping materials to modify semiconductors to work effectively in the visible range. The bandgap between the valance and conduction bands is reduced when the semiconductor material has been altered with doping elements. As doping materials, several transition metals such as V, Co, Mn, W, Fe, Cr, Fe, and Mo have been widely employed [53]. A few scientists have investigated the co-doping technique in the semiconductor framework using two metals or two nonmetals. It has been discovered that co-doping greatly improves the quantum efficiency of photocatalytic processes by reducing the rate of electron-hole pair recombination. This mixed phase can also aid electron and whole charge transfer between rutile and anatase TiO_2 forms.

Single-doped and co-doped photocatalysts can break down organic pollutants. Co-doped semiconductor catalysts are typically more effective than single-doped semiconductors for the degradation of organic wastewater. The doping element significantly modifies the electron structure of semiconductors. The modified semiconductors allow photons with lower energy to excite, generating hydroxyl and superoxide radicals under visible light, which non-selectively degrade organic pollutants [58]. In addition, dopants can boost the quantum efficiency of TiO_2 photocatalytic processes by preventing the generated electron-hole pairs from recombining [60]. Dopants, especially double metal/non-metal dopants, have been intensively explored because of their synergistic influence on photocatalyst visible-light absorption efficiency. It can slow down the rate of photo-induced carrier recombination. Co-doped TiO_2 had higher photocatalytic activity than single-doped TiO_2 [53].

7.4.6 IRON-BASED PHOTO NANOCATALYST

Iron, the fourth most prevalent metal in the Earth's crust, is a comparatively harmless element [59]. Iron possesses a range of valences, including 0, +2, and +3, and it has a diverse set of properties. Iron-based materials/catalysts are widely used in catalytic reactions. The present section summarizes the iron-based photocatalyst for the degradation of organic wastewater. Photocatalysts with narrow band gaps are appealing now a days. The semiconducting characteristics of iron based catalysts have been widely used to degrade organic wastewater [60]. An iron-based photocatalyst has attracted particular attention because of its ease of availability, non-toxicity, and narrow bandgap (for example, 2.2 eV). These photocatalysts exhibit easier excitation than other metal oxides [61]. Heo et al. [62] developed a heterojunction photocatalyst (CdS/Fe_2O_3) for the photoreduction of CO_2. Yang et al. [61] monitored rhodamine pollutant degradation using Fe-heterojunction (g-C_3N_4/ZnO@Fe_3O_4) and the Fe-organic ligand catalyst (2-amino-5-fluorobenzotrifluoride@Fe_3O_4). Sunlight-responsive catalysts based on iron have a promising future for photocatalytic degradation of organic wastewater.

7.5 CONCLUSION

The many potential strategies for fabricating Photo-nanocatalyst are summarized in this chapter. Photocatalytic degradation under visible light is regarded as one of the most sustainable technologies for treating organic wastewater. Hence, the degradation of organic wastewater using Photo-nanocatalyst has garnered considerable attention. Numerous parameters that significantly affect nano-based photocatalytic degradation have been reviewed. Photo catalytic degradation of organic pollutants are also summerised in order to know the effectiveness of the photocatalytic process. Most photocatalytic reactions have only been studied on a laboratory basis. As a result, a pilot-scale investigation is essential to the determination of the method's sustainability.

REFERENCES

[1] Mishra, D. and Srivastava, M. "Low-dimensional nanomaterials for the photocatalytic degradation of organic pollutants." *Nanomaterials as Photocatalysts for Degradation of Environmental Pollutants: Challenges and Possibilities* (2020) 15–38. doi: 10.1016/B978-0-12-818598-8.00002-X.

[2] Zhu, G. F. Li, J. Z. Wu, P. Jin, H. Z. and Wang, Z. "The performance and phase separated characteristics of an anaerobic baffled reactor treating soybean protein processing waste water." *Bioresource Technology* 99 (2008) 8027–8033. doi: 10.1016/J.BIORTECH.2008.03.046.

[3] Bertanza, G, Collivignarelli, C. and Pedrazzani, R. The role of chemical oxidation in combined chemical-physical and biological processes: experiences of industrial waste water treatment, Water Science and Technology, 44(5), 109-116, (2001).

[4] Mensah, K. A. and Forster, C. F. "An examination of the effects of detergents on anaerobic digestion." *Bioresource Technology* 90 (2003) 133–138. doi: 10.1016/S0960-8524(03)00126-3.

[5] Kim, K. H. and Ihm, S. K. "Heterogeneous catalytic wet air oxidation of refractory organic pollutants in industrial waste waters: a review." *Journal of Hazardous Materials* 186 (2011) 16–34. doi: 10.1016/J.JHAZMAT.2010.11.011.

[6] Barrabés, N. and Sá, J. "Catalytic nitrate removal from water, past, present and future perspectives." *Applied Catalysis B: Environmental* 104 (2011) 1–5. doi: 10.1016/j.apcatb.2011.03.011.

[7] Lu, K. Q. Li, Y. H. Zhang, F. Qi, M. Y. Chen, X. Tang, Z. R. Yamada, Y. M.A. Anpo, M. Conte, M. and Xu, Y. J. "Rationally designed transition metal hydroxide nanosheet arrays on graphene for artificial CO_2 reduction." *Nature Communications* 11 (2020) doi: 10.1038/s41467-020-18944-1.

[8] Fujishima, A. and Honda, K. "Electrochemical photolysis of water at a semiconductor electrode." Nature, 238(5358), 37-38, (1972).

[9] Tahir, M. B. Rafique, M. Rafique, M. S. Nawaz, T. Rizwan, M. and Tanveer, M. "Photocatalytic nanomaterials for degradation of organic pollutants and heavy metals." *Nanotechnology and Photocatalysis for Environmental Applications* (2020) 119–138. doi: 10.1016/B978-0-12-821192-2.00008-5.

[10] Sapkota, B. B. and Mishra, S. R. "A simple ball milling method for the preparation of p-CuO/n-ZnO nanocomposite photocatalysts with high photocatalytic activity."

Journal of Nanoscience and Nanotechnology 13 (2013) 6588–6596. doi: 10.1166/jnn.2013.7544.

[11] Behnajady, M. A. Eskandarloo, H. Modirshahla, N. and Shokri, M. "Investigation of the effect of sol–gel synthesis variables on structural and photocatalytic properties of TiO_2 nanoparticles." *Desalination* 278 (2011) 10–17. doi: 10.1016/J.DESAL.2011.04.019.

[12] Ahmed, M. A. El-Katori, E. E. and Gharni, Z. H. "Photocatalytic degradation of methylene blue dye using Fe_2O_3/TiO_2 nanoparticles prepared by sol–gel method." *Journal of Alloys and Compounds* 553 (2013) 19–29. doi: 10.1016/J.JALLCOM.2012.10.038.

[13] Mizukoshi, Y. Makise, Y. Shuto, T. Hu, J. Tominaga, A. Shironita, S. and Tanabe, S. "Immobilization of noble metal nanoparticles on the surface of TiO_2 by the sonochemical method: photocatalytic production of hydrogen from an aqueous solution of ethanol." *Ultrasonics Sonochemistry* 14 (2007) 387–392. doi: 10.1016/j.ultsonch.2006.08.001.

[14] Behnajady, M. A. and Eskandarloo, H. "Silver and copper co-impregnated onto TiO_2-p25 nanoparticles and its photocatalytic activity." *Chemical Engineering Journal* 228 (2013) 1207–1213. doi: 10.1016/j.cej.2013.04.110.

[15] Zulfiqar, M. Chowdhury, S. and Omar, A. A. "Hydrothermal synthesis of multiwalled TiO_2 nanotubes and its photocatalytic activities for orange ii removal." *Separation Science and Technology (Philadelphia)* 53 (2018) 1412–1422. doi: 10.1080/01496395.2018.1444050.

[16] Muñoz-Fernandez, L. Sierra-Fernandez, A. Milošević, O. and Rabanal, M. E. "Solvothermal synthesis of Ag/ZnO and Pt/ZnO nanocomposites and comparison of their photocatalytic behaviors on dyes degradation." *Advanced Powder Technology* 27 (2016) 983–993. doi: 10.1016/J.APT.2016.03.021.

[17] Sun, H. Wang, C. Pang, S. Li, X. Tao, Y. Tang, H. and Liu, M. "Photocatalytic TiO_2 films prepared by chemical vapor deposition at atmosphere pressure." *Journal of Non-Crystalline Solids* 354 (2008) 1440–1443. doi: 10.1016/j.jnoncrysol.2007.01.108.

[18] Thirumal, V. Yuvakkumar, R. Kumar, P. S. Ravi, G. Keerthana, S. P. and Velauthapillai, D. "Facile single-step synthesis of MXene@CNTs hybrid nanocomposite by CVD method to remove hazardous pollutants." *Chemosphere* 286, 131733, (2022). doi: 10.1016/j.chemosphere.2021.131733.

[19] Venkatachalam, N. Palanichamy, M. and Murugesan, V. "Sol–gel preparation and characterization of alkaline earth metal doped nano TiO_2: efficient photocatalytic degradation of 4-chlorophenol." *Journal of Molecular Catalysis A: Chemical* 273 (2007) 177–185. doi: 10.1016/J.MOLCATA.2007.03.077.

[20] Malengreaux, C. M. Pirard, S. L. Bartlett, J. R. and Heinrichs, B. "Kinetic study of 4-nitrophenol photocatalytic degradation over a Zn^{2+} doped TiO_2 catalyst prepared through an environmentally friendly aqueous sol-gel process." *Chemical Engineering Journal* 245 (2014) 180–190. doi: 10.1016/j.cej.2014.01.068.

[21] Jaiswal, R. Patel, N. Dashora, A. Fernandes, R. Yadav, M. Edla, R. Varma, R. S. Kothari, D. C. Ahuja, B. L. and Miotello, A. "Efficient co-b-codoped tio2 photocatalyst for degradation of organic water pollutant under visible light." *Applied Catalysis B: Environmental* 183 (2016) 242–253. doi: 10.1016/j.apcatb.2015.10.041.

[22] Salavati-Niasari, M. Soofivand, F. Sobhani-Nasab, A. Shakouri-Arani, M. Faal, A.Y. and Bagheri, S. "Synthesis, characterization, and morphological control of zntio3 nanoparticles through sol-gel processes and its photocatalyst application." *Advanced Powder Technology* 27 (2016) 2066–2075. doi: 10.1016/J.APT.2016.07.018.

[23] Hayat, K. Gondal, M. A. Khaled, M. M. and Ahmed, S. "Effect of operational key parameters on photocatalytic degradation of phenol using nano nickel oxide synthesized by sol-gel method." *Journal of Molecular Catalysis A: Chemical* 336 (2011) 64–71. doi: 10.1016/j.molcata.2010.12.011.

[24] Khade, G. V. Suwarnkar, M. B. Gavade, N. L. and Garadkar, K. M. "Sol–gel microwave assisted synthesis of sm-doped TiO_2 nanoparticles and their photocatalytic activity for the degradation of methyl orange under sunlight." *Journal of Materials Science: Materials in Electronics* 27 (2016) 6425–6432. doi: 10.1007/s10854-016-4581-7.

[25] Balcha, A. Yadav, O. P. and Dey, T. "Photocatalytic degradation of methylene blue dye by zinc oxide nanoparticles obtained from precipitation and sol-gel methods." *Environmental Science and Pollution Research* 23 (2016) 25485–25493. doi: 10.1007/s11356-016-7750-6.

[26] Isai, K. A. and Shrivastava, V. S. "Photocatalytic degradation of methylene blue using ZnO and 2%Fe–ZnO semiconductor nanomaterials synthesized by sol–gel method: a comparative study." *SN Applied Sciences* 1, 438-443 and 1247, (2019). doi: 10.1007/s42452-019-1279-5.

[27] Thilagavathi, P. Manikandan, A. Sujatha, S. Jaganathan, S. K. and Antony, S. A. "Sol–gel synthesis and characterization studies of Nimoo4 nanostructures for photocatalytic degradation of methylene blue dye." *Nanoscience and Nanotechnology Letters* 8 (2016) 438–443. doi: 10.1166/nnl.2016.2150.

[28] Soltani, T. and Lee, B. K. "Novel and facile synthesis of Ba-doped $BiFeO_3$ nanoparticles and enhancement of their magnetic and photocatalytic activities for complete degradation of benzene in aqueous solution." *Journal of Hazardous Materials* 316 (2016) 122–133. doi: 10.1016/j.jhazmat.2016.03.052.

[29] Rajaitha, P. M. Hajra, S. Sahu, M. Mistewicz, K. Toroń, B. Abolhassani, R. Panda, S. Mishra, Y. K. and Kim, H. J. "Unraveling highly efficient nanomaterial photocatalyst for pollutant removal: a comprehensive review and future progress." *Materials Today Chemistry* 23 (2022) 100692. doi: 10.1016/J.MTCHEM.2021.100692.

[30] Riaz, N. Chong, F. K. Dutta, B. K. Man, Z. B. Khan, M. S. and Nurlaela, E. Effect of calcination temperature on Orange II photocatalytic degradation using Cu:Ni/TiO2 under visible light, in: 2011 *Natl. Postgrad. Conf.—Energy Sustain. Explor. Innov. Minds* (2011). doi: 10.1109/NatPC.2011.6136258.

[31] Singh, J. Sharma, S. Aanchal, and Basu, S. "Synthesis of Fe_2O_3/TiO_2 monoliths for the enhanced degradation of industrial dye and pesticide via photo-fenton catalysis." *Journal of Photochemistry and Photobiology A: Chemistry* 376 (2019) 32–42. doi: 10.1016/J.JPHOTOCHEM.2019.03.004.

[32] A. M. A. Khalil, E. M. Farag, A. B. El-Fadl, M. M. A. and El-Aassar, "Photocatalytic degradation of organic pollutants in waste water using different nanomaterials immobilized on polymeric beads." *Desalination and Water Treatment* 193 (2020) 117–128. doi: 10.5004/dwt.2020.25680.

[33] Lei, P. Wang, F. Gao, X. Ding, Y. Zhang, S. Zhao, J. Liu, S. and Yang, M. "Immobilization of TiO_2 nanoparticles in polymeric substrates by chemical bonding for multi-cycle photodegradation of organic pollutants." *Journal of Hazardous Materials* 227–228 (2012) 185–194. doi: 10.1016/J.JHAZMAT.2012.05.029.

[34] Prabhakar V. Veerendra, S. Byon, C. and Reddy, C. V. "ZrO_2/MoS_2 heterojunction photocatalysts for efficient photocatalytic degradation of methyl orange." *Electronic Materials Letters* 12 (2016) 812–823. doi: 10.1007/s13391-016-6267-y.

[35] Hu, L. Deng, G. Lu, W. Pang, S. and Hu, X. "Deposition of CdS nanoparticles on MIL-53(Fe) metal-organic framework with enhanced photocatalytic degradation of RhB under visible light irradiation." *Applied Surface Science* 410 (2017) 401–413. doi: 10.1016/J.APSUSC.2017.03.140.

[36] Li, J. Xie, Z. Xiong, Y. Li, Z. Huang, Q. Zhang, S. Zhou, J. Liu, R. Gao, X. Chen, C. Tong, L. Zhang, J. and Liu, Z. "Architecture of β-graphdiyne-containing thin film using modified glaser–hay coupling reaction for enhanced photocatalytic property of TiO_2." *Advanced Materials* 29 (2017) doi: 10.1002/adma.201700421.

[37] Thangavel, S. Krishnamoorthy, K. Krishnaswamy, V. Raju, N. Kim, S. J. and Venugopal, G. "Graphdiyne-ZnO nanohybrids as an advanced photocatalytic material." *Journal of Physical Chemistry C* 119 (2015) 22057–22065. doi: 10.1021/acs.jpcc.5b06138.

[38] Yan, Y. Yang, H. Yi, Z. Li, R. and Wang, X. "Enhanced photocatalytic performance and mechanism of $Au@CaTiO_3$ composites with au nanoparticles assembled on $CaTiO_3$ nanocuboids." *Micromachines* 10, 254, (2019). doi: 10.3390/mi10040254.

[39] Jiang, W. Li, Z. Liu, C. Wang, D. Yan, G. Liu, B. and Che, G. "Enhanced visible-light-induced photocatalytic degradation of tetracycline using BiOI/MIL-125(Ti) composite photocatalyst." *Journal of Alloys and Compounds* 854, 157166 (2021). doi: 10.1016/j.jallcom.2020.157166.

[40] Zhang, Z. Li, X. Liu, B. Zhao, Q. and Chen, G. "Hexagonal microspindle of NH_2-MIL-101(Fe) metal-organic frameworks with visible-light-induced photocatalytic activity for the degradation of toluene." *RSC Advances* 6 (2016) 4289–4295. doi: 10.1039/c5ra23154j.

[41] Wu, G. Li, P. Xu, D. Luo, B. Hong, Y. Shi, W. and Liu, C. "Hydrothermal synthesis and visible-light-driven photocatalytic degradation for tetracycline of Mn-doped $SrTiO_3$ nanocubes." *Applied Surface Science* 333 (2015) 39–47. doi: 10.1016/j.apsusc.2015.02.008.

[42] Cao, M. Wang, P. Ao, Y. Wang, C. Hou, J. and Qian, J. "Photocatalytic degradation of tetrabromobisphenol a by a magnetically separable graphene-TiO_2 composite photocatalyst: mechanism and intermediates analysis." *Chemical Engineering Journal* 264 (2015) 113–124. doi: 10.1016/j.cej.2014.10.011.

[43] Zhang, W. Xiao, X. Zheng, L. and Wan, C. "Fabrication of TiO_2/MoS_2 composite photocatalyst and its photocatalytic mechanism for degradation of methyl orange under visible light." *Canadian Journal of Chemical Engineering* 93 (2015) 1594–1602. doi: 10.1002/cjce.22245.

[44] Ramos-Delgado, N. A. Gracia-Pinilla, M. A. Maya-Treviño, L. Hinojosa-Reyes, L. Guzman-Mar, J. L. and Hernández-Ramírez, A. "Solar photocatalytic activity of TiO_2 modified with WO_3 on the degradation of an organophosphorus pesticide." *Journal of Hazardous Materials* 263 (2013) 36–44. doi: 10.1016/j.jhazmat.2013.07.058.

[45] Tang, Y. Dong, L. Mao, S. Gu, H. Malkoske, T. and Chen, B. "Enhanced photocatalytic removal of tetrabromobisphenol a by magnetic coo@graphene nanocomposites under visible-light irradiation." *ACS Applied Energy Materials* 1 (2018) 2698–2708. doi: 10.1021/acsaem.8b00379.

[46] Payan, A. Fattahi, M. Jorfi, S. Roozbehani, B. and Payan, S. "Synthesis and characterization of titanate nanotube/single-walled carbon nanotube (TNT/SWCNT) porous nanocomposite and its photocatalytic activity on 4-chlorophenol degradation under UV and solar irradiation." *Applied Surface Science* 434 (2018) 336–350. doi: 10.1016/j.apsusc.2017.10.149.

[47] Wang, S. L. Zhu, H. W. Tang, W. H. and Li, P. G. "Propeller-shaped ZnO nanostructures obtained by chemical vapor deposition: photoluminescence and photocatalytic properties." *Journal of Nanomaterials* 2012 (2012) doi: 10.1155/2012/594290.

[48] Niu, H. Yang, Y. Zhao, W. Lv, H. Zhang, H. and Cai, Y. "Single-crystalline Fe_7S_8/Fe_3O_4 coated zero-valent iron synthesized with vacuum chemical vapor deposition technique: enhanced reductive, oxidative and photocatalytic activity for water purification." *Journal of Hazardous Materials* 401 (2021) 123442. doi: 10.1016/J.JHAZMAT.2020.123442.

[49] Subramaniam, M. N. Goh, P. S. Lau, W. J. Ismail, A. F. Gürsoy, M. and Karaman, M. "Synthesis of titania nanotubes/polyaniline via rotating bed-plasma enhanced chemical vapor deposition for enhanced visible light photodegradation." *Applied Surface Science* 484 (2019) 740–750. doi: 10.1016/j.apsusc.2019.04.118.

[50] Xiao, G. Zeng, H. Y. Xu, S. Chen, C. R. Zhao, Q. and Liu, X. J. "Preparation of Ti species coating hydrotalcite by chemical vapor deposition for photodegradation of azo dye." *Journal of Environmental Sciences (China)* 60 (2017) 14–23. doi: 10.1016/j.jes.2017.03.031.

[51] Bento, R. T. Correa, O. V. and Pillis, M. F. "Photocatalytic activity of undoped and sulfur-doped TiO_2 films grown by MOCVD for water treatment under visible light." *Journal of the European Ceramic Society* 39 (2019) 3498–3504. doi: 10.1016/J.JEURCERAMSOC.2019.02.046.

[52] Zhang, Y. Hawboldt, K. Zhang, L. Lu, J. Chang, L. and Dwyer, A. "Carbonaceous nanomaterial-TiO_2 heterojunctions for visible-light-driven photocatalytic degradation of aqueous organic pollutants." *Applied Catalysis A: General* 630 (2022) 118460. doi: 10.1016/J.APCATA.2021.118460.

[53] Akerdi, A. G. and Bahrami, S. H. "Application of heterogeneous nano-semiconductors for photocatalytic advanced oxidation of organic compounds: a review." *Journal of Environmental Chemical Engineering* 7 (2019) 103283. doi: 10.1016/j.jece.2019.103283.

[54] Meda, U. S. Vora, K. Athreya, Y. and Mandi, U. A. "Titanium dioxide based heterogeneous and heterojunction photocatalysts for pollution control applications in the construction industry." *Process Safety and Environmental Protection*, 161, 771-787 (2022). doi: 10.1016/J.PSEP.2022.03.066.

[55] Wu, Y. Deqin Meng, Qingbin Guo, Dengzheng Gao, and Li Wang. "Study on TiO_2/g-C_3N_4 s-scheme heterojunction photocatalyst for enhanced formaldehyde decomposition." *Optical Materials* 126 (2022) 112213. doi: 10.1016/J.OPTMAT.2022.112213.

[56] Zhang, X. Wang, J. Dong, X. X. and Lv, Y. K. "Functionalized metal-organic frameworks for photocatalytic degradation of organic pollutants in environment." *Chemosphere* 242 (2020) 125144. doi: 10.1016/J.CHEMOSPHERE.2019.125144.

[57] Li, S. Shan, S. Chen, S. Li, H. Li, Z. Liang, Y. Fei, J. Xie, L. and Li, J. "Photocatalytic degradation of hazardous organic pollutants in water by Fe-MOFs and their composites: a review." *Journal of Environmental Chemical Engineering* 9 (2021) 105967. doi: 10.1016/J.JECE.2021.105967.

[58] Chen, D. Cheng, Y. Zhou, N. Chen, P. Wang, Y. Li, K. Huo, S. Cheng, P. Peng, P. Zhang, R. Wang, L. Liu, H. Liu, Y. and Ruan, R. "Photocatalytic degradation of organic pollutants using TiO_2-based photocatalysts: a review." *Journal of Cleaner Production* 268 (2020) 121725. doi: 10.1016/J.JCLEPRO.2020.121725.

[59] Tanaka, S. Kaneti, Y. V. Septiani, N. L. W. Dou, S. X. Bando, Y. Hossain, M. S. A. Kim, J. and Yamauchi, Y. "A review on iron oxide-based nanoarchitectures for biomedical, energy storage, and environmental applications." *Small Methods* 3, 1800512, (2019). doi: 10.1002/smtd.201800512.

[60] Liu, X. Ma, R. Zhuang, L. Hu, B. Chen, J. Liu, X. and Wang, X. "Recent developments of doped g-C3N4 photocatalysts for the degradation of organic pollutants." *Critical Reviews in Environmental Science and Technology* 51 (2021) 751–790. doi: 10.1080/10643389.2020.1734433.

[61] Yang, S. Sun, Q. Shen, Y. Hong, Y. Tu, X. Chen, Y. and Zheng, H. "Design, synthesis and application of new iron-based cockscomb-like photocatalyst for high effectively degrading water contaminant under sunlight." *Applied Surface Science* 525 (2020) 146559. doi: 10.1016/J.APSUSC.2020.146559.

[62] Heo, J. N. Shin, J. Do, J. Y. Kim, R. and Kang, M. "Reliable carbon dioxide photoreduction by a rational electron transfer cycle formed on a nanorod-shaped CDS/Fe_2O_3 heterojunction catalyst." *Applied Surface Science* 495, 143567 (2019). doi: 10.1016/j.apsusc.2019.143567.

8 Recent Development in Nanofiltration Membrane for Water Purification

Sayed Zenab Hasan,[1] Sapna Nehra,[2]
Rekha Sharma,[3] and Dinesh Kumar[4*]*
[1]Department of Chemistry Dr. K. N. Modi University, Newai, Rajasthan, India
[2]Department of Chemistry, Nirwan University, Jaipur, Rajasthan, India, sharma20rekha@gmail.com
[3]Department of Chemistry, Banasthali Vidyapith, Banasthali, Rajasthan, India
[4]School of Chemical Sciences, Central University of Gujarat, Gandhinagar, 382030, India
*Corresponding authors:
Email: nehrasapna111@gmail.com, dinesh.kumar@cug.ac.in

CONTENTS

8.1 Introduction 147
8.2 Nanofiltration Membranes for Water Purification 151
    8.2.1 Polymeric Nanofiltration Membranes 151
    8.2.2 Ceramic Nanofiltration Membranes 153
        8.2.2.1 Metal Oxide Nanofiltration Membranes 155
        8.2.2.2 Carbon-based Nanofiltration Membranes 156
    8.2.3 Mixed Matrix Nanofiltration Membranes 158
8.3 Conclusion..159
References..160

8.1 INTRODUCTION

The fast expansion of world population and the decline of water quality in the environment has produced a serious risk to the increasing human population [1]. To solve this problem, many techniques have been explored for the optimum use of water resources. Various techniques, such as photodegradation, adsorption, membrane separation, biodegradation, electrocoagulation, Fenton - assisted photo-oxidation, and the like, are available to reduce pollutants. Sometimes, two combinations of two or more of these technologies are used [2]. Among these methods, membrane technologies are extensively examined in water treatment because they are sustainable and energy efficient. Nanofiltration (NF) has acquired vast recognition as a promising membrane separation technology due to its versatility. Its pore size in the range of 0.5 - 2 nm has separation ability and great small molecule permeability. NF membranes are widely used in organic separation and desalination [3]. NF is a kind of pressure-driven filtration technology which has properties within ultrafiltration (UF) and reverse osmosis (RO). NF has a more significant rejection of various inorganic and organic substances than UF, due to its surface charge property and pore size.

The rejection of polyvalent inorganic salts is high in NF at modest pressures. These properties make NF more competitive in cost-benefit and selectivity than RO. Therefore, the progress of NF has gained recognition and popularity globally in recent years [4]. Compared to RO membranes, NF membranes have a looser polymer network that requires lower operating pressures, typically less than 25 bar. Because of this characteristic, the NF process has been used for water softening, desalination of brackish water, food processing, industrial waste water, and waste water reuse [5].

The general process of constructing a thin-film composite (TFC) NF membrane is interfacial polymerization (IP). In which acid chloride and organic amines irreversibly combine at the interface of two immiscible phases (namely, the organic phase and the aqueous phase) for synthesizing a layer of polyamide (PA) having a thickness of about ≤ 200 nm. NF membranes produced in this way could reach higher rejections of most dissolved organic molecules and multivalent iron salts based on electrostatic interaction and pore size exclusion [6].

Most of the investigators have dedicated studies for the fabrication of NF membranes with regular pore sizes or by regulating the charges or hydrophilicity onto the selected layer for improving the membrane's selectivity. Some advancement has been gained in the reduction of membrane fouling. Zhang et al. [7] comprehensively explain the recent approach of developing NF membrane selectivity (Figure 8.1), focused on membrane fabrication methods, pore size, pore–size distribution changing, charge distribution control, and surface antifouling alteration.

TFC polyamide membranes have been examined and applied industrially. Still, the restrictions of high fouling tendency and flux-selectivity limit their NF procedure applications. Table 8.1 shows the challenges and research requirements in these membranes. Hence, the need for improved membrane techniques and the exploration of novel materials are required [8].

This chapter imparts a thorough general outlook on the recent approach for developing NF membrane selectivity (Figure 8.2). NF membranes contribute wide areas of application, but in this chapter, the focus is on NF membranes for water purification.

Nanofiltration Membrane for Water Purification

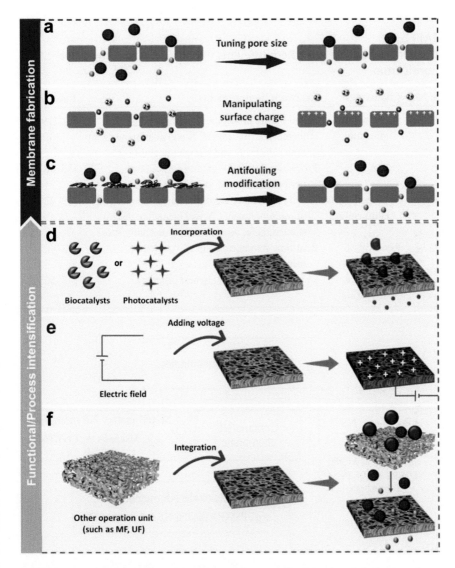

FIGURE 8.1 Scheme for improving NF membranes selectivity (a - c) during membrane fabrication or (d - f) function or procedure intensification (Adapted with permission from [7]. Copyright @2020, American Chemical Society).

TABLE 8.1
Challenges and Research Requirements in NF Membrane For Water Purification

S. No.	Challenges	Research Requirements	References
1	High energy utilization	Novel hybrid procedures and low-energy membrane procedures. Novel membranes with high selectivity and high permeability	[6], [19], [33, 34]
2	Fouling	Procedure design for greater antifouling performance. Antifouling membranes	[35–38]
3	Micropollutants elimination	Novel membranes are modified for the removal of micropollutants. Novel procedure design (for example, osmotic membrane bioreactor)	[39–42]
4	Removal of pathogens	Real-time monitoring and affirmation of membrane integrity	[43–46]
5	Concentrate discarding	Development of novel hybrid process	[47–50]

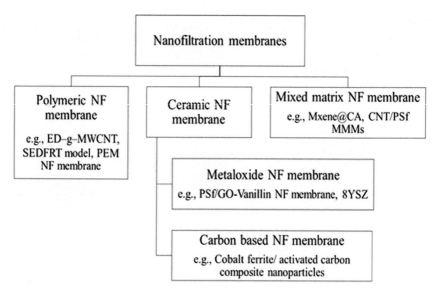

FIGURE 8.2 Classification of nanofiltration membranes for water purification based on recent developments.

8.2 NANOFILTRATION MEMBRANES FOR WATER PURIFICATION

8.2.1 Polymeric Nanofiltration Membranes

Pedayash et al. [9] developed a hybrid NF membrane by joining a positive surface charge using a self-assembly of ethylenediamine (ED) to MWCNTs (ED-g-MWCNTs) by coating asymmetric polyethersulfone (PES). The effect of nanotube stacking on the layer, isoelectric point, surface roughness, hydrophilicity, as well as mechanical and thermal stability was evaluated. In addition, pure water fluxes of the synthesized membranes increased by 122% compared to the pure PES membrane, reaching 80.5 L/m^{2}h. Because of the low surface roughness and high surface hydrophilicity, the antifouling properties of membranes were enhanced remarkably. The entire estimation for the membranes, reviewing every major factor, was achieved by categorizing the efficiency of products. Outcomes showed the integration of nanotubes by a loading ratio of 0.6 wt.% thus indicating a useful membrane. Ultimately, the outcomes of this study introduced a solution for the purification of water from heavy metal ions formulated on PES NF membrane integrated with ED–g–MWCNT.

Meschke et al. [10] regulated NF separation by tuning the supply and the membrane characteristics. In the analysis, the aqueous solution included important elements: rhenium (Re), molybdenum (Mo), cobalt (Co), and germanium (Ge) with a broad scale of concentrations and differing ionic species. The supply corresponds to a bioleachate of German flue dust arising from the smelting of copper ore. The separation performance of the eight economic polymeric NF membranes, estimated in various cross-flow experiments, and the impact of transmembrane flow velocity, pressure, flow regime, ionic strength, and recovery were explored. It was noted that an enhanced ionic strength induced by an increased zinc (Zn) concentration decreased permeability and influenced Re retention negatively. The separation selectivity towards Re is intensified by increasing the trans-membrane pressure.

Hu et al. [11] proposed a facile coordination driven assembly procedure to establish an interlayer of multifunctional polysodium 4–styrenesulfonate (PSS) metal ion complex onto the substrate to construct a highly efficient TFC NF membrane having creased morphology. The PSS metal ion interlayer functions as a nanoscaffold to give an even IP platform, which is beneficial for setting up an accurate and thinner polyamide layer with less interfered polyamide in the pores of the support. It behaves as a macromolecule in influencing the diffusion-driven instability of the IP method, which brings about the formation of a creased polyamide layer with enhanced surface area. The TFC membrane that was produced exhibited a four-fold permeance compared to the conventional TFC–0 membrane and sustained a competitive rejection for Na_2SO_4. Due to decreased effective pore size and increased surface negative charges, the novel NF membrane showed low retention of divalent cations (Mg/Ca) and high rejection of trace organic compounds in tap water, showing significant ability to treat drinking water.

The NF treatment procedure could be optimized to valorize two-phase olive–oil washing waste water. Its phenolic fraction recovery and gaining of a refined permeate stream could also be optimized. Ochando–Pulido et al. [12] used a factorial design to optimize the procedure. Outcomes were elucidated with the help of the response surface technique. A statistical multifactorial investigation was carried out to evaluate every conjugated effect of a capable complex of the NF process input parameters. Then the procedure was eventually designed by an equation of the second–grade quadratic fitting model. Ultimately, the parametric attribute standards were examined. The suggested olive oil washing waste water purifying procedures were 26.5 bar operating pressure, 32.7 m/s tangential velocities, 35°C system temperature, and 3.7 pH, confirming stable and high membrane flux of 106.2 L/h/m². This optimized data was very suitable for the highly capable NF membrane TFC PA/polysulfone (PSf) at raw effluent pH and ambient temperature conditions. The advanced conditions produced a permeate stream that could be recycled for irrigation and concentrated about 6.5 times in volume, with 1315.7 mg/L of entire phenolic content.

Lopez et al. [13] developed an algorithm combining the solution electro diffusion model, dissolved species homogeneous reactions, and concentration polarization solution electro diffusion film integrated with reactive transport (SEDFRT model) and executed it to design the electrolyte mixtures species NF. This produced workable evaluations of an average thickness of the concentration polarization layer, together with the species' membrane permeances. The membrane penetrability for species sustained a similar sequence of magnitude. Besides solutions comprising two trace salts: $NaBr$, NH_4Cl, and dominant salts $MgSO_4/MgCl_2$. The two trace monovalent cations, NH_4^+ and Na^+, proceed simply through the membrane described by a mixture of dielectric and Donnan exclusion.

It was noticed that $MgCl_2$ fractions, NH_4^+ and Na^+ were sped up at large supply conditions by emerging electric fields in the membrane. However, more supply of $MgSO_4$ mole fractions developed enhanced single charge cation rejections because of depletion in these fields. Eventually, the study expanded the design of solute transport in NF membranes to consider equilibrium reactions among species and, hence, the development of complexes. The membrane penetrability of the associated species was evaluated for the first time. A related situation was recognized in a solution of multi electrolytes imitating the configuration of surface waters influenced by urban, agriculture, and industrial pollution, where about five solutes, carbonate, calcium, sulfate, ammonium, and magnesium, were recognized as potentially influenced by equilibrium complexation reactions.

Elsho et al. [14] systematically studied the extended pH stability of polyelectrolyte multilayer (PEM) NF membranes which displayed potential under severe pH conditions. They also studied four polyelectrolyte systems exhibited ex-situ to 0.1 M $NaOH$ and 0.1 M HNO_3 solutions afterward. This work was assessed by using a selected mixture of weak and strong polyelectrolytes. Stability and membrane behaviors differ, which can be accounted for by the electrostatic relation strength between polyelectrolytes in exposure conditions. Hence, it could be deduced that poly–diallyl–dimethyl–ammonium–chloride (PDADMAC)/PSS PEM NF membranes showed good behavior under extreme pH conditions. Their performance was still stable after over two months of exposure. Further, poly–allylamine–hydrochloride (PAH)/PSS PEM NF membranes were stable in an adverse acidic environment. Both membranes displayed stable performances when subjected to more excessive pH conditions. All of these can use PEM NF membranes, particularly PDADMAC/PSS, in extreme pH environments in application areas, for example, mining, textile, and dairy industries.

Liu et al. [15] considered porous organic polymers to be favorable as a model of highly efficient membranes, mostly due to better surface area, built-in mesoporous organization, and great polymer potential. Stratified o–hydroxyazo porous organic polymers (o–POPs) beside piperazine (PIP) monomers were included in the casting solution, then phase inversion to synthesize PIP, incorporating polymeric substrates. They fabricated o–POP altered thin-film nanocomposite (TFN) membranes using IP and trimesoyl chloride. The integrated o–POPs can enhance PIP in substrates and prevent its diffusion for the organic interface in IP reactions by a physical barrier and hydrogen bonding, accompanying the constitution of the crumpled membrane surface. Hence, an increment of hydrophilicity and surface roughness provided TFN membranes with notably enhanced water permeability and sustained a solute rejection. An ideal TFN membrane with an o–POP amount of 0.3 wt.% showed impressive water permeance of 29.6 L/m^2/h/bar, a high 94.9% rejection of Na_2SO_4, and an orderly elimination of reactive dyes, which formed o–POP TFN membranes competitive for NF in the purification of water.

8.2.2 Ceramic Nanofiltration Membranes

Bethi et al. [16] studied hydrodynamic cavitation integrated with ceramic NF membranes to eliminate organic pollutants from waste water from the textile industry. Hydrodynamic cavitation integrated with ceramic NF membrane and dropwise incorporation of 2 mL/L hydrogen peroxide showed the removal of 36.62% of total organic carbon. Hydrodynamic cavitation /Fenton/ceramic NF membrane and dropwise addition of H_2O_2 showed 58.85% of total organic carbon elimination and the waste water was nearly colorless in 3 hrs. Incorporating H_2O_2 also remarkably increased the efficiency of the process by diminishing total organic carbon and color. A collective effect was noticed when mixing hydrodynamic cavitation, Fenton and NF. On a reuse-discharge basis, the treated waste water was not appropriate for reuse and discharge because the characteristics of the final waste water detected from the analysis were total organic carbon (TOC) of 471 mg/L and chemical oxygen demand (COD) of 847 mg/L. Therefore, this suggested treating the RO system further regarding the discharge and reuse criteria.

Kramer et al. [17] used ceramic NF membranes to treat municipal sewage. Although, the understanding of fouling management techniques for this application is limited. Frequent used fouling control procedures, chemical cleaning, had disadvantages. Chemical cleaning, mainly sodium hypochlorite, to eliminate organic fouling, negatively affects the tubular ceramic NF membranes' glass seal layers and the surroundings. For that reason, the usage of chemical cleaning had to be confined to a degree.

To begin with, the common fouling control techniques for polymeric UF and microfiltration (MF) membranes relied on ceramic NF membranes. A precoat process integrated with a chemical reaction, assisted the disconnection of established cake coating. Before the initiation of filtration, a precoat film was filtered over the membrane surface. Then the pre-coated layer reacted as a boundary between the membrane surface and foulants. After that filtration, the precoat layer reacted with the cleaning reagent below the fouling layer, allowing rapid fouling elimination. The outcome was that hydraulic backwash was unsuccessful in this kind of membrane. However, the forward flush helped sustain a higher flux with more relative downtime production. Hence, the reaction-based precoat was very potent in sustaining a high flux and emerged in the highest overall water production. Therefore, from the two reactions tested based on precoat procedures, the reaction of a citric acid with calcium carbonate was more successful than a Fenton reaction.

Cabrera et al. [18] used a commercial titania ceramic NF membrane unit to use a 20 m^3/h permeate flow capacity to lower ion concentration, TOC, and total suspended solids (TSS) in reused water in a Canadian oil sand mine. In addition, the study was examined for about two years to assess membrane performance in actual recycling water conditions. Hence, the study aimed at a 50% stage cut outcome. The strong interconnection between rejection and specific flux with the highest mass rejections was noticed at the least specific flux values. The development of a potential cake layer onto the surface of the membrane appeared to favor the rejection as low specific flux values enhanced mass rejection. Also, the investigation of around 20 ions displayed those differences in electrostatic phenomena, and hydrated ionic sizes with divalent cations showed the largest rejection. Further TOC of 75 - 90% and nearly 100% rejection of TSS were noticed. The outcomes showed that applying the technique in an oil sand mine is practicable, obtaining remarkable improvements in water quality and minimizing river water consumption.

Agtas et al. [19] performed work with real wastewater collected from disperse and reactive printing. Further, target its treatment and recovery of hot water. Therefore, this used a pilot-scale ceramic UF and NF membrane system. In ten cycles, NF and UF membranes were used, and only NF membranes were used in the last four cycles. In addition, all the samples were put through the total hardness, COD, TOC, conductivity, and color tests. Temperature and flux were observed during operations. Hence, in the achieved outcomes, for both UF and NF cycles altogether, average removal efficiencies were 89%, 86.4%, 83.5%, and 68% for COD, TOC, color, and color hardness, respectively.

Natural organic matter (NOM) can be successfully removed by treating drinking water from surface water using ion exchange. The formation of spent ion exchange regeneration brine, a contaminant costly to address, was the major

Nanofiltration Membrane for Water Purification

drawback of utilizing ion exchange to remove NOM. Caltran et al. [20] considered ceramic NF to remedy this consumed NOM rich brine, having a target to reduce waste and salt recycling content. It has a benefit compared to polymeric NF in that fouling is finite. When NOM is concentrated and rejected, a clear permeate with the regeneration salt NaCl can form and recycle again in the ion exchange regeneration procedure. NOM can beneficially be removed from NaCl solution using steric effects; disclosed by bench-scale studies. Although removing NaCl from different salts existing in the brine, such as Na_2SO_4, did not meet recycling objectives. Also, the lower sulfate rejection was chiefly because of the membrane's low zeta potential at the high brine ionic strength.

8.2.2.1 Metal Oxide Nanofiltration Membranes

Agrawal et al. [21] used membrane technology for essential waste water treatment processes for several decades because of its selectivity, permeability, and removal ability. Biofouling produces serious problems with membrane permeability, and reduces membrane life and selectivity. In contrast, polymeric membranes are extensively used for waste water treatment because of their higher flexibility, great pore forming potential, and economy. They are more susceptible to fouling and restricted by their hydrophobicity property. Hence, metal oxide nanomaterials have been effectively used to develop polymer nanocomposite membranes because of their pore channels, hydrophilicity, and greater surface areas. These act as antibacterial agents used to produce polymer nanocomposite membranes, also having antibacterial properties and anti-biofouling characteristics.

Yadav et al. [22] reported a sequence of PSf/ graphite oxide (GO) –vanillin NF membranes that were selective, fouling resistant, and highly permeable. The membranes are created by 2D GO layers inserted with vanillin and PSf as the base polymer. There is increasing attention to the combined effect of vanillin and GO on enhancing antifouling characteristics and the permeability of membranes. They utilized microscopic and spectroscopic techniques to obtain complete morphological and physicochemical studies. This developed $PSf_{16}/GO_{0.15}$–vanillin$_{0.8}$ membrane showed 25.4% and 92.5% rejection rates for 2000 ppm NaCl and $MgSO_4$ solutions. The outcomes of antifouling showed over 99% rejection for bovine serum albumin and a 93.57% flux recovery ratio. Outcomes displayed 84 - 90% rejection towards Mg(II) and Ca(II) along with a 90.32% flux recovery ratio. Hence, the analysis experimentally showed that combining vanillin and GO with the polymeric matrix improved membrane performance and fouling resistance.

Laminar structure membranes fabricated with 2D substances, such as graphene oxide, influenced researchers' consideration in waste water treatment. Despite that, there is still a difficulty in getting high-performance membranes by accurately regulating the channel of water permeability. Deng et al. [23] employed a sequence of graphene oxide nanosheets and silica nanoparticles (SiO_2 NPs) to regulate the channel microstructure of GO/SiO_2 composite NF membranes. Utilizing small-sized nanosheets of graphene oxide in the composite membrane can decrease the length of the transverse channel in the middle of GO nanosheets and enhance the amount of longitudinal pores along the edges of GO nanosheet, increasing the composite membrane water flux. Incorporating small-sized SiO_2 NPs within the composite membrane can help set up a continuous transverse channel of water permeability and significantly increase water flux. Hence, outcomes displayed that the composite membrane developed with GO of 1.2 μm and SiO_2 NPs of 25 nm and 30 wt.% of the GO mass had a great performance with 72.8 L/m²/h pure water flux and dyes rejection rate greater than 99.4%. After being recycled eight times, the dyes rejection rate was 97.5% high. Hence, the study displayed that regulating the size of NPs and nanosheets can successfully regulate the channel microstructure of composite membranes, having importance for membrane performance.

Qin et al. [24] prepared yttria–stabilized ZrO_2 (8YSZ) (8 mol%) NF membranes from size–regulated spherical ZrO_2 NPs along with approximately 10 nm average diameter and a particle roundness value of over 0.90, with a reverse micelle–mediated sol-gel procedure successfully constructing the NPs. The doping of yttria repressed the tetragonal to a monoclinic phase transition, established the membrane's integrity, enlarged the specific surface area, reduced the tetragonal size of the grain, contracted the distribution of pore size, and hence enhanced the performance of NF. The developed 8YSZ NF membranes of approximately 260 nm of thickness showed high performances of NF, whereas the molecular weight cut–off and pure water permeability were 800 ± 50 Da and 3.9 - 4.2 L/m²/h/bar, respectively. The elimination rate of carbofuran through developing 8YSZ NF membranes was above 82% in pesticide waste water treatment, whereas the maximum elimination rate obtained was 89%. In addition, the used membranes can always be renewed after low–temperature calcination and alkali wash, suitable for various reuses.

8.2.2.2 Carbon-based Nanofiltration Membranes

Undoubtedly, carbon-based NF membranes are promising candidates for advanced progress in membrane technology. Sonawanea et al. [25] aimed to summarize and feature the latest growth in carbon-based NF membranes to purify water. The presented membranes have a great capability for the elimination of arsenic, heavy metals, NOM, self care products, pharmaceuticals, and the inactivation of viruses and bacteria in waste water. In addition, these membranes have the potential to treat oily waste water. However, carbon-based materials have excellent properties that exhibit their capabilities for application in water purification. Fundamental and applied work is necessary to address issues such as durability, uniform dispersion, the arrangement of carbon-based material, complete contact, the extensive manufacture of membranes, and high value, with stability, all of which are difficult tasks. CNT–based polymer composite membrane has particular advantages for treating water over other membranes. The better water permeability, high mechanical stability, solute rejection capability, good ion resistance, and antifouling nature, are valuable properties for the advancement of these membranes.

Zhao et al. [26] prepared a novel alkynyl carbon composite NF membrane based on polydopamine, polyethyleneimine, and alkynyl carbon material coprecipitation.

Alkynyl carbon material is a graphite derivative that comprises sp^1 and sp^2 hybrid carbons and has a uniform pore distribution of 3–5 nm. In addition, it is ideal for selective and highly permeable membranes because of its high porosity, extreme thinness, and superhydrophobic surface. X-ray photoelectron spectroscopy (XPS), Fourier transform infrared spectroscopy (FTIR), and scanning electron microscopy (SEM) validated that alkynyl carbon material was beneficially combined on the polyacrylonitrile substrate surface. The potential zeta study displayed that the membrane surface is super hydrophilic because of pore structures and superhydrophobicity. The NF experiments displayed that the membrane water flux was 366 L/m^2/h/MPa, which yielded a great rejection of NOM and organic dyes in water. For example, the rejection and permeation flux of humic acid and methyl blue are 94% and 285 L/m^2/h/MPa and 96% and 360 L/m^2/h/MPa, respectively.

The membrane exhibited strong running stability. Hence, the advanced membrane separation material has good prospects for potential application in waste water treatment and water purification. Alkynyl carbon materials are easy to prepare and cheaper than carbon nanotubes and graphene oxide.

Zareei et al. [27] prepared a mixed matrix PES–based NF membrane with integrating cobalt ferrite/activated carbon composite NPs. These composite NPs were developed utilizing the chemical precipitation method. The field emission scanning electron microscope (FESEM), X-ray diffraction (XRD), and FTIR were utilized to distinguish the developed NPs. The produced membranes were analyzed by SEM, mean pore size, 3D surface image, water contact angle, and porosity estimation. Antifouling ability, salt rejection, and water flux of membranes were studied. The combined membranes displayed a smoother and hydrophilic surface compared with the pristine membrane. The Na_2SO_4 and NaCl rejections were improved from 55% to 95% and 42% to 85% using cobalt ferrite/activated carbon NPs in the membrane structure. The combined membranes showed an improved antifouling property and highly stable flux compared with the pristine membrane. The highest Cu^{2+}, Pb^{2+}, and Ni^{2+} rejections were 97%, 86%, and 90%, respectively, for the combined membrane having one wt.% cobalt ferrite/activated carbon as compared to others.

Song et al. [28] demonstrated the viability of eliminating the sequence of cationic small-sized pollutants such as dyes, hardness-causing ions, and pharmaceutically active compounds (PhACs) in an aqueous system with TFN membranes. The effects on IP for nanostructured quaternized carbon-based nanoparticles (QCBNPs), characteristics involving wettability, smoothness, and positive charge, were remarkably enhanced after doping alterations. These modifications were achieved by enhancing each small-sized pollutant's elimination rate, fouling resistance, and membrane permeability. Particularly, the optimal TFN–quaternized carbon quantum dots membrane had high permeate flux of 23.8 L/m^2/h/bar, and flux rejection rates for $MgCl_2$ 98.4%, cationic trimethoprim 99.7%, atenolol 99.5%, rhodamine B 98.9%, and methylene blue 98.4%, flux recovery ratio of 85.5%, that was enhanced by 29.6%, 53.6%, 6.75%, 5.96%, 2.17%, 2.21%, and 17.0% respectively, as compared to the tidy membrane.

8.2.3 Mixed Matrix Nanofiltration Membranes

Pandey et al. [29] manufactured a new mixed–matrix NF membrane that was fouling–resistant related to covalently cross-linked $Ti_3C_2T_x$ (MXene)/cellulose acetate (MXene@CA) using phase inversion and then by formaldehyde cross-linking. The developed MXene@CA membranes' physicochemical properties were establishe through water contact angle, FESEM, XRD, and energy dispersive spectroscopy (EDS). The performance of the developed membranes was estimated to define the rejection properties, bacterial growth reluctance, and water flux. The 10% MXene@CA membrane showed greater than 98%, and 92% rejection of methyl green and rhodamine B. 10% MXene@CA membrane showed greater than 96% and 98% growth reluctance for B. subtilis and E. coli. This developed NF membrane, mainly 10% MXene@CA, could be advised for biomedical applications and water purification.

Bandehali et al. [30] performed a surface functionalization of GO nanoplates by glycidyl polyhedral oligosilsesquioxane (POSS) to create polyetherimide related NF membranes using the phase inversion process. The developed membranes were distinguished with FTIR, FESEM, atomic force microscopy (AFM), and XRD. The result of various amounts of glycidyl POSS–GO (PG) in the membrane matrix on the antifouling property, and separation performance of combined membranes was examined using pure water flux, contact angle measurement, flux recovery ratio, Na_2SO_4, $Cu(NO_3)_2$, $CrSO_4$, and $Pb(NO_3)_2$. The combined polyetherimide/PG membranes displayed crucial antifouling properties and partition performance as compared with pure polyetherimide/graphene oxide and polyetherimide membranes. The combined polyetherimide (PG) membranes displayed excellent rejection as compared to combined PG/GO membranes and pure polyetherimide. Further, the highest flux recovery ratio (FRR) rate was discovered at 96% for the combined membrane, while the FRR was 40% for polyetherimide/graphene oxide and 33% for pure polyetherimide membranes.

Ferreira et al. [31] prepared CNT/PSf mixed matrix membranes (MMMs) by the non–solvent influenced phase inversion process and implemented in the partition of lignin from black liquor and developed three kinds of membranes to reduce fouling, particularly the main problem concerning the implementation of polymeric membranes for black liquor partitioning. The result of CNT on morphological properties and filtration was explored by examining mean pore size, porosity, pure water flux, SEM, AFM, FTIR, and water contact angle. These studies showed that with 0 - 0.5% enhanced nanotubes content there was an increase of mean pore size approximately three times, at 15 bar 14.8 - 132.5 $L/m^2/h$ water flux, 74.4 - 79.7% porosity, and limiting of the surface coarseness of developed membranes. In addition, this work on membranes was examined concerning fouling resistance and lignin rejection by utilizing standard lignoboost lignin solution at different pressures compared with a commercial PES membrane. Hence, it was concluded that permeate flux depletion was least, approximately 10%, at a pressure of 12 bar for 0.5% CNT-based membranes. Although, regarding the lignin rejection, the lowest (52 - 76%) and highest (80 - 98%) rates were attained with the 0.1% CNT membrane and 0%

CNT PSf membrane, respectively. Besides, fouling data displayed that with using chemical cleaning, the starting flux was fully retrieved for CNT-based membranes without interrupting their rejection behavior and flux. Hence, 0.5% CNT membranes showed a great balance among lignin rejections and permeated flux between the three kinds of developed membranes. In conclusion, the applications of developed membranes for industrial black liquor NF, pre-treated by UF, permit the reaching of 37 - 42% rejection for hemicelluloses and 82 - 84% for lignin.

Low filler loading, thick separation layers, and boundary imperfections limit the performance of MMMs between filler and polymer. Shu et al. [32] synthesized and designed a new 2D metal-organic framework (2D MOF), called BUT-203. The micron-sized crystals of this MOF can be quickly divided into layers of nanosheets with around 3 nm thicknesses and a lateral size of hundreds of nanometers of different microns. These nanosheets had a great affinity with the polycation polymer polyethyleneimine in ethanol. For anionic dye molecules, the membrane showed water permeance of $870 \, L/m^2/h/MPa$ and about 97.9% rejections. In conclusion, the membrane presented extended running stability, great antifouling property, and high dye desalination ability.

8.3 CONCLUSION

A huge amount of work has been done to address the scarcity of clean water, and nanotechnology is an efficient technology with rapid progress. The commercialization and study of NF membranes emerged in the early 1960s. Two types of membranes: TFC membranes and cellulose-based membranes, affect water purification. Advance investigation routes for the barrier surface in TFC membranes involve advancement in fouling resistance, besides thermal and chemical stabilities. Concurrently, microporous supports could be improved to enhance permeability and mechanical strength. Since the 1980s, inorganic NF membranes have been analyzed on a lab scale for purifying water. Carbon-based and metal oxide ceramics are illustrative. The chief blending process for the metal oxide membranes is a sol-gel method that requires more development to manage particle distribution and size. The work of MMMs produced with both inorganic and organic nanomaterials is great but they are still very costly compared to the other membranes. Thus, this is crucial to get the economic conflict of MMMs, and capable applications. Hence, nanotechnology directs a path to improve NF membranes in water purification. Still, scientific and technical issues must be resolved before more advantages can be seen. Regardless of the challenges, it is extremely probable that ceramic membranes will be industrialized and commercialized for the purification of water and for desalination fields. The latest advancement in 3D printing techniques for membrane production is a great activity for modern development in membrane science or for good research to be managed. Membrane technology is a rising field, however, these recent challenges of membrane stability, long–term stability, high flux permeation, industrial-scale development, reusability, and commercial and practical approaches to membrane manufacturing need to be addreessed and further research is required.

REFERENCES

[1] Bai, L. Liu, Y. Ding, A. Ren, N. Li, G. and Liang, H. "Fabrication and characterization of thin-film composite (TFC) nanofiltration membranes incorporated with cellulose nanocrystals (CNCs) for enhanced desalination performance and dye removal." *Chemical Engineering Journal* 358 (2019), 1519–1528. doi:10.1016/j.cej.2018.10.147.

[2] Thombre, N. V. Gadhekar, A. P. Patwardhan, A. V. and Gogate, P. R. "Ultrasound induced cleaning of polymeric nanofiltration membranes." *Ultrasonics Sonochemistry* 62 (2020), 104891. doi:10.1016/j.ultsonch.2019.104891.

[3] Guo, Y-S. Ji, Y-L. Wu, B. Wang, N-X. Yin, M-J. An, Q-F. and Gao, C-J. "High-flux zwitterionic nanofiltration membrane constructed by in-situ introduction method for monovalent salt/antibiotics separation." *Journal of Membrane Science* 593 (2020), 117441. doi:10.1016/j.memsci.2019.117441.

[4] Borjigin, B. Yu, L. Xu, L. Zhao, C. and Wang, J. "Influence of incorporating beta zeolite nanoparticles on water permeability and ion selectivity of polyamide nanofiltration membranes." *Journal of Environmental Sciences* 98 (2020), 77–84. doi:10.1016/j.jes.2020.04.010.

[5] Wang, R. and Lin, S. "Pore model for nanofiltration: History, theoretical framework, key predictions, limitations, and prospects." *Journal of Membrane Science* 620 (2021), 118809. doi:10.1016/j.memsci.2020.118809.

[6] Qin, D. Huang, G. Terada, D. Jiang, H. Ito, M. M. Gibbons, A. H. Igarashi R. et al. "Nanodiamond mediated interfacial polymerization for high performance nanofiltration membrane." *Journal of Membrane Science* 603 (2020), 118003. doi:10.1016/j.memsci.2020.118003.

[7] Zhang, H. He, Q. Luo, J. Wan, Y. and Darling, S. B. "Sharpening nanofiltration: strategies for enhanced membrane selectivity." *ACS Applied Materials & Interfaces* 12, no. 36 (2020), 39948–39966. doi:10.1021/acsami.0c11136.

[8] Sun, H. and Wu, P. "Tuning the functional groups of carbon quantum dots in thin film nanocomposite membranes for nanofiltration." *Journal of Membrane Science* 564 (2018), 394–403. doi:10.1016/j.memsci.2018.07.044.

[9] Peydayesh, M. Mohammadi, T. and Nikouzad, S. K. "A positively charged composite loose nanofiltration membrane for water purification from heavy metals." *Journal of Membrane Science* 611 (2020), 118205. doi:10.1016/j.memsci.2020.118205.

[10] Meschke, K. Hansen, N. Hofmann, R. Haseneder, R. and Repke, J-U. "Influence of process parameters on separation performance of strategic elements by polymeric nanofiltration membranes." *Separation and Purification Technology* 235 (2020), 116186. doi:10.1016/j.seppur.2019.116186.

[11] Hu, P. Tian, B. Xu, Z. and Niu, Q. J. "Fabrication of high performance nanofiltration membrane on a coordination-driven assembled interlayer for water purification." *Separation and Purification Technology* 235 (2020), 116192. doi:10.1016/j.seppur.2019.116192.

[12] Ochando-Pulido, J. M. Corpas-Martínez, J. R. Vellido-Perez, J. A. and Martinez-Ferez, A. "Optimization of polymeric nanofiltration performance for olive-oil-washing waste water phenols recovery and reclamation." *Separation and Purification Technology* 236 (2020), 116261. doi:10.1016/j.seppur.2019.116261.

[13] López, J. Yaroshchuk, A. Reig, M. Gibert, O. and Cortina, J. L. "An engineering model for solute transport in semi-aromatic polymeric nanofiltration membranes: Extension of Solution-Electro-Diffusion model to complex mixtures." *Journal of Environmental Chemical Engineering* 9, no. 4 (2021), 105262. doi:10.1016/j.jece.2021.105262.

[14] Elshof, M. G. de Vos, W. M. de Grooth, J. and Benes, N. E. "On the long-term pH stability of polyelectrolyte multilayer nanofiltration membranes." *Journal of Membrane Science* 615 (2020), 118532. doi:10.1016/j.memsci.2020.118532.

[15] Liu, Y. Zhu, J. Zheng, J. Gao, X. Tian, M. Wang, X. Xie, Y. F. Zhang, Y. Volodin, A. and Van der Bruggen, B. "Porous organic polymer embedded thin-film nanocomposite membranes for enhanced nanofiltration performance." *Journal of Membrane Science* 602 (2020), 117982. doi:10.1016/j.memsci.2020.117982.

[16] Bethi, B. Sonawane, S. H. Bhanvase, B. A. and Sonawane, S. S. "Textile industry waste water treatment by cavitation combined with fenton and ceramic nanofiltration membrane." *Chemical Engineering and Processing-Process Intensification* 168 (2021), 108540. doi:10.1016/j.cep.2021.108540.

[17] Kramer, F. C. Shang, R. Rietveld, L. C. and Heijman, S. J. G. "Fouling control in ceramic nanofiltration membranes during municipal sewage treatment." *Separation and Purification Technology* 237 (2020), 116373. doi:10.1016/j.seppur.2019.116373.

[18] Cabrera, S. M. Winnubst, L. Richter, H. Voigt, I. and Nijmeijer, A. "Industrial application of ceramic nanofiltration membranes for water treatment in oil sands mines." *Separation and Purification Technology* 256 (2021), 117821. doi:10.1016/j.seppur.2020.117821.

[19] Agtaş, M. Yılmaz, Ö. Dilaver, M. Alp, K. and Koyuncu, I. "Hot water recovery and reuse in textile sector with pilot scale ceramic ultrafiltration/nanofiltration membrane system." *Journal of Cleaner Production* 256 (2020), 120359. doi:10.1016/j.jclepro.2020.120359.

[20] Caltran, I. Rietveld, L. C. Shorney-Darby, H. L. and Heijman, S. G. J. "Separating NOM from salts in ion exchange brine with ceramic nanofiltration." *Water Research* 179 (2020), 115894. doi:10.1016/j.watres.2020.115894.

[21] Agrawal, A. Sharma, A. Awasthi, K. K. and Awasthi, A. "Metal oxides nanocomposite membrane for biofouling mitigation in waste water treatment." *Materials Today Chemistry* 21 (2021), 100532. doi:10.1016/j.mtchem.2021.100532.

[22] Yadav, S. Ibrar, I. Samal, A. K. Altaee, A. Déon, S. Zhou, J. and Ghaffour, N. "Preparation of fouling resistant and highly perm-selective novel PSf/GO-vanillin nanofiltration membrane for efficient water purification." *Journal of Hazardous Materials* 421 (2022), 126744. doi:10.1016/j.jhazmat.2021.126744.

[23] Deng, H. Zheng, Q. Chen, H. Huang, J. Yan, H. Ma, M. Xia, M. Pei, K. Ni, H. and Ye, P. "Graphene oxide/silica composite nanofiltration membrane: adjustment of the channel of water permeation." *Separation and Purification Technology* 278 (2021), 119440. doi:10.1016/j.seppur.2021.119440.

[24] Qin, H. Guo, W. Huang, X. Gao, P. and Xiao, H. "Preparation of yttria-stabilized ZrO_2 nanofiltration membrane by reverse micelles-mediated sol-gel process and its application in pesticide waste water treatment." *Journal of the European Ceramic Society* 40, no. 1 (2020), 145–154. doi:10.1016/j.jeurceramsoc.2019.09.023.

[25] Sonawane, S. L. Labhane, P. K. and Sonawane, G. H. "Carbon-based nanocomposite membranes for water purification." In *Handbook of Nanomaterials for Waste*

water Treatment, edited by Bharat Bhanvase, Shirish Sonawane, Vijay Pawade, Aniruddha Pandit eBook ISBN: 9780128214992 pp. 555–574. Elsevier (2021). doi:10.1016/B978-0-12-821496-1.00036-2.

[26] Zhao, G. Wang, X. Li, C. and Meng, H. "Superhydrophilic alkynyl carbon composite nanofiltration membrane for water purification." *Applied Surface Science* 508 (2020), 144788. doi:10.1016/j.apsusc.2019.144788.

[27] Zareei, F. Bandehali, S. Parvizian, F. Hosseini, S. M. and Shen, J. N. "Promoting the separation and antifouling properties of polyethersulfone-based nanofiltration membrane by incorporating of cobalt ferrite/activated carbon composite nanoparticles." *Chemical Engineering Research and Design* 169 (2021), 204–213. doi:10.1016/j.cherd.2021.03.016.

[28] Song, Y. Wang, Y. Zhang, N. Li, X. Bai, X. and Li, T. "Quaternized carbon-based nanoparticles embedded positively charged composite membranes towards efficient removal of cationic small-sized contaminants." *Journal of Membrane Science* 630 (2021), 119332. doi:10.1016/j.memsci.2021.119332.

[29] Pandey, R. P. Rasheed, P. A. Gomez, T. Azam, R. S. and Mahmoud, K. A. "A fouling-resistant mixed-matrix nanofiltration membrane based on covalently cross-linked $Ti_3C_2T_x$ (MXene)/cellulose acetate." *Journal of Membrane Science* 607 (2020), 118139. doi:10.1016/j.memsci.2020.118139.

[30] Bandehali, S. Moghadassi, A. Parvizian, F. Zhang, Y. Hosseini, S. M. and Shen, J. "New mixed matrix PEI nanofiltration membrane decorated by glycidyl-POSS functionalized graphene oxide nanoplates with enhanced separation and antifouling behaviour: heavy metal ions removal." *Separation and Purification Technology* 242 (2020), 116745. doi:10.1016/j.seppur.2020.116745.

[31] Ferreira, I. Alves, P. Gil, M. H. and Gando-Ferreira, L. M. "Lignin separation from black liquor by mixed matrix polysulfone nanofiltration membrane filled with multiwalled carbon nanotubes." *Separation and Purification Technology* 260 (2021), 118231. doi:10.1016/j.seppur.2020.118231.

[32] Shu, L. Xie, L-H. Meng, Y. Liu, T. Zhao, C. and Li, J-R. "A thin and high loading two-dimensional MOF nanosheet based mixed-matrix membrane for high permeance nanofiltration." *Journal of Membrane Science* 603 (2020), 118049. doi:10.1016/j.memsci.2020.118049.

[33] Chen, K. Li, P. Zhang, H. Sun, H. Yang, X. Yao, D. Pang, X. Han, X. and Niu, Q. J. "Organic solvent nanofiltration membrane with improved permeability by in-situ growth of metal-organic frameworks interlayer on the surface of polyimide substrate." *Separation and Purification Technology* 251 (2020), 117387. doi:10.1016/j.seppur.2020.117387.

[34] Liu, Z. Wang, T. Wang, D. and Mi, Z. "Regulating the morphology of nanofiltration membrane by thermally induced inorganic salt crystals for efficient water purification." *Journal of Membrane Science* 617 (2021), 118645. doi:10.1016/j.memsci.2020.118645.

[35] Mi, Y-F. Xu, G. Guo, Y-S. Wu, B. and An, Q-F. "Development of antifouling nanofiltration membrane with zwitterionic functionalized monomer for efficient dye/salt selective separation." *Journal of Membrane Science* 601 (2020), 117795. doi:10.1016/j.memsci.2019.117795.

[36] An, X. Zhang, K. Wang, Z. Ly, Q. V. Hu, Y. and Liu, C. "Improving the water permeability and antifouling property of the nanofiltration membrane grafted with

hyperbranched polyglycerol." *Journal of Membrane Science* 612 (2020), 118417. doi:10.1016/j.memsci.2020.118417.

[37] Ren, L. Chen, J. Lu, Q. Han, J. and Wu, H. "Antifouling nanofiltration membrane fabrication via surface assembling light-responsive and regenerable functional layer." *ACS Applied Materials & Interfaces* 12, no. 46 (2020), 52050–52058. doi:10.1021/acsami.0c16858.

[38] Izadmehr, N. Mansourpanah, Y. Ulbricht, M. Rahimpour, A. and Omidkhah, M. R. "TETA-anchored graphene oxide enhanced polyamide thin film nanofiltration membrane for water purification; performance and antifouling properties." *Journal of Environmental Management* 276 (2020), 111299. doi:10.1016/j.jenvman.2020.111299.

[39] Wang, P. Wang, F. Jiang, H. Zhang, Y. Zhao, M. Xiong, R. and Ma, J. "Strong improvement of nanofiltration performance on micropollutant removal and reduction of membrane fouling by hydrolyzed-aluminum nanoparticles." *Water Research* 175 (2020), 115649. doi:10.1016/j.watres.2020.115649.

[40] Cuhorka, J. Wallace, E. and Mikulasek, P. "Removal of micropollutants from water by commercially available nanofiltration membranes." *Science of the Total Environment* 720 (2020), 137474. doi:10.1016/j.scitotenv.2020.137474.

[41] Zhang, H. Luo, J. Woodley, J. M. and Wan, Y. "Confining the motion of enzymes in nanofiltration membrane for efficient and stable removal of micropollutants." *Chemical Engineering Journal* 421 (2021), 127870. doi:10.1016/j.cej.2020.127870.

[42] Dai, R. Wang, X. Tang, C. Y. and Wang, Z. "Dually charged MOF-based thin-film nanocomposite nanofiltration membrane for enhanced removal of charged pharmaceutically active compounds." *Environmental Science & Technology* 54, no. 12 (2020), 7619–7628. doi:10.1021/acs.est.0c00832.

[43] Park, M. and Snyder, S. A. "Attenuation of contaminants of emerging concerns by nanofiltration membrane: Rejection mechanism and application in water reuse." In *Contaminants of Emerging Concern in Water and Waste water*, edited by Arturo Hernandez-Maldonado, Lee Blaney pp. 177–206. Butterworth-Heinemann (2020), doi:10.1016/B978-0-12-813561-7.00006-7

[44] Barro, L. Nebie, O. Chen, M-S. Wu, Y-W. Koh, M. B. C. Knutson, F. Watanabe, N. Takahara, M. and Burnouf, T. "Nanofiltration of growth media supplemented with human platelet lysates for pathogen-safe xeno-free expansion of mesenchymal stromal cells." *Cytotherapy* 22, no. 8 (2020), 458–472. doi:10.1016/j.jcyt.2020.04.099.

[45] Goswami, K. P. and Pugazhenthi, G. "Credibility of polymeric and ceramic membrane filtration in the removal of bacteria and virus from water: A review." *Journal of environmental management* 268 (2020), 110583. doi:10.1016/j.jenvman.2020.110583.

[46] Xu, S. Lu, D. Qi, J. Wang, P. Zhao, Y. Zhang, H. and Ma, J. "Gravity-driven multifunctional microporous membranes for household water treatment: simultaneous pathogenic disinfection, metal recycling, and biofouling mitigation." *Chemical Engineering Journal* 410 (2021), 128289. doi:10.1016/j.cej.2020.128289.

[47] Lin, J. Chen, Q. Liu, R. Ye, W. Luis, P. Van der Bruggen, B. and Zhao, S. "Sustainable management of landfill leachate concentrate via nanofiltration

enhanced by one-step rapid assembly of metal-organic coordination complexes." *Water Research* 204 (2021), 117633. doi:10.1016/j.watres.2021.117633.

[48] Arola, K. Mänttäri, M. and Kallioinen, M. "Two-stage nanofiltration for purification of membrane bioreactor treated municipal waste water–Minimization of concentrate volume and simultaneous recovery of phosphorus." *Separation and Purification Technology* 256 (2021), 117255. doi:10.1016/j.seppur.2020.117255.

[49] Li, K. Liu, Q. Fang, F. Wu, X. Xin, J. Sun, S. Wei, Y. et al. "Influence of nanofiltration concentrate recirculation on performance and economic feasibility of a pilot-scale membrane bioreactor-nanofiltration hybrid process for textile waste water treatment with high water recovery." *Journal of Cleaner Production* 261 (2020), 121067. doi:10.1016/j.jclepro.2020.121067.

[50] Du, X. Li, Z. Xiao, M. Mo, Z. Wang, Z. Li, X. and Yang, Y. "An electro-oxidation reactor for treatment of nanofiltration concentrate towards zero liquid discharge." *Science of the Total Environment* 783 (2021), 146990. doi:10.1016/j.scitotenv.2021.146990.

9 Nanomaterials for Electrochemical Treatment of Pollutants in Water

Chetan Kumar[1] and Ritu Painuli[2*]
[1]Natural Product Chemistry Division, CSIR–Indian Institute of Integrative Medicine, Canal Road, Jammu 180001, India, kumarbelwal@gmail.com
[2]Department of Chemistry, Banasthali Vidyapith, Banasthali, Tonk 304022, India
*Corresponding author: Email: ritsjune8.h@gmail.com

CONTENTS

9.1 Introduction 165
9.2 Detection of Nutrients and Phenolic Compounds 166
    9.2.1 Detection of Nitrates 167
    9.2.2 Detection of Phosphates 167
    9.2.3 Detection of Phenolic Compounds 168
9.3 Detection of Heavy Metal (HMs) 168
    9.3.1 Carbon Nanotubes 170
    9.3.2 Carbon Nanotubes/Gold Nanoparticles 170
    9.3.3 Bismuth Nanoparticles 171
    9.3.4 Iron Oxides 172
9.4 Detection of Chemical Oxygen Demand 173
    9.4.1 Copper Nanoparticles 173
    9.4.2 Graphene Oxide (GO)/Copper Nanoparticles 175
9.5 Conclusion 176
Acknowledgments 176
References 177

9.1 INTRODUCTION

The socio-economic growth encouraged by the exponential rise in the human population has globally reduced the water quality [1-3]. Due to industrial waste, insufficient sewage treatment, radioactive waste materials, and the like, water

DOI: 10.1201/9781003252931-9

contamination has resulted in the deposition of pollutants into the ecosystem [4,5]. Water pollution damages the environment and can also be accountable for air pollution that causes dangerous results for human beings. The emerging pollutants that require widespread attention include heavy metals such as mercury, lead, and chemical pollutants [6]. Although heavy metals are obtained from biogeochemical mechanisms, some heavy metals in the aquatic systems are derived from fossil fuel combustion, the release of municipal waste water, and the mining processes [7, 8]. Personal care products and the disruptive endocrine chemicals from industrial and pharmaceutical applications are the foremost components of evolving pollutants [9 - 12]. Heavy metal ions have damaging impacts on the environment and the health of humans because of their toxic nature and their ability to accumulate along with the trophic systems [13]. These pollutants have adverse effects on human beings, causing, for instance, disturbances in hormonal systems and damage to body organs [14, 15]. Therefore, there is an urgent need to protect ecosystems via the identification of contaminants. Currently used analytical techniques, for instance, gas chromatography, high-performance liquid chromatography, mass spectroscopy, and so forth, are precise, reliable, and sensitive. These have various shortcomings such as expense, difficulties with typical sample preparation, their use of harmful solvents, and the requirement for trained operators [16 - 18]. These methods become inadequate for on-site and in situ analysis.

Conversely, electrochemical approaches have been widely functional for the on-site detection of toxic pollutants because of their merits such as their low detection limits, their superior sensitivity, their cost-effective nature, their rapid analytical response, their portability, and their simple preparation [19 - 21]. The present chapter discusses the modern advancement in electrochemical devices for detecting pollutants in waste water. The sensing of various different pollutants, namely, heavy metals, organic matter, nutrients, and phenolic compounds, is addressed. Several core issues will also be explored associated with this rapidly evolving area that need to be considered for further research.

9.2 DETECTION OF NUTRIENTS AND PHENOLIC COMPOUNDS

The most hazardous pollutants of groundwater across the globe include nitrates, phosphates, and ammonia. Extensive utilization of these pollutants in products such as cosmetics, furniture, paint, and the like, has been responsible for water pollution. The increased concentration of these nutrients can cause serious concerns for human well-being, leading to ailments such as diabetes, liver damage, kidney damage, and cancer. They have been responsible for O_2 reduction in water bodies and damaging wildlife [22–24]. In addition, phenolic compounds, for instance, bisphenol A (BPA), catechol (CA), phenol (phe), hydroquinne (HQ), are also evolving as contaminants because of their nonbiodegradable nature, which has a lethal effect on the environment, on animals and on humans.

9.2.1 DETECTION OF NITRATES

Among the various electrodes used for nitrate detection, copper has emerged as the most employed electrode owing to its short overpotential and fast response for the reduction of nitrate ions [25]. Copper nanowire arrays were effortlessly developed on Cu wires by the electrochemical reduction of CuO nanowires made by thermal oxidation. The nanostructured seamless surfaces presented enhanced conductivity and improved surface area. The prepared sensor detected nitrate ions over 50 - 600 μM, and a LOD of 12.2 μM was achieved on the nanostructured copper wire [26]. Nanostructured sensor-based Cu nanowire arrays were prepared via the galvanic deposition method for the electrochemical detection of nitrate ions. The prepared sensor displays shot response time with a LOD of less than 10 μM. The ready sensor was also tested for the real water samples such as drinking water, rainwater, and river water [27]. Cu electrodes were fabricated via thermal annealing of Cu nanowires at 600 - 800 °C (in an Ar atmosphere) for the electrochemical sensing of nitrate ions. The electrochemically active surfaces of the Cu electrodes showed that the active surface areas of the electrodes decreased on increasing the annealing temperature. It was also observed that an electrode fabricated at 600°C displayed maximum nitrate reduction under acidic conditions. The sensor showed a linear response range of 8 - 5860 μM with a LOD of 1 - 35 μM [28]. The electrochemical reduction capability of a graphene-modified Cu electrode (GMCE) and a bare Cu electrode (BCE) were studied. It was noticed that the electrocatalytic performance of the prepared GMCE showed 3.5-fold enhancement in reduction peak densities as of BCE. The amphoteric response of the modified electrode was enhanced as a function of nitrate concentration with a LOD of 10 μM and linear range from 9.0×10^{-6} to 9.4×10^{-4} M and a fast response for nitrate reduction [29].

9.2.2 DETECTION OF PHOSPHATES

Cobalt (Co), a hard metal, has received consideration as the cobalt oxide (CoO) group has shown satisfactory selectivity for the various phosphate species such as PO_4^{3-}, $H_2PO_4^{-}$, HPO_4^{2-} [30]. A novel sensor based on the electrodepositing of Co–Fe(II) alloy on a chip with gold as a working electrode was developed to determine dihydrogen phosphate ($H_2PO_4^{-}$). The prepared sensor displayed a fast response time and was stable for approximately 16 days. It also displayed a linear response from 10^{-6} M to 10^{-2} M with an LOD of 3.41×10^{-5} M. To improve the sensor efficacy, the actual chip was heated to 400°C. The improved sensor showed a detection limit of 2.14×10^{-6} M and a range from 10^{-6} M to 1×10^{-1} M [31]. Wang et al. reported a nano–cobalt sensor made using a hydrothermally grown ZnO nanoflake template and using an electroplated method for phosphate sensing. With the help of the electroplating method, ZnO nanoflakes were covered via the Co layer and were etched to prepare the final nanostructured Co electrodes. The developed Co electrode displayed a LOD from 1×10^{-6} to 1×10^{-3} M [32]. By electroplating Co on the screen-printing electrode surface, a potentiometric Co-based screen printing sensor was prepared to determine $H_2PO_4^{2-}$. The prepared sensor displayed a response from 10^{-5} mol/L to 10^{-1} mol/L with an LOD of 3.16×10^{-6} in acidic solution [33].

9.2.3 Detection of Phenolic Compounds

AgNPs and AuNPs possess inimitable optical, catalytic, and electrical properties and have been extensively employed for sensing applications [34 - 36]. Goulart and coworkers reported that a glassy carbon electrode (GCE) modified by MWCNTs and AgNPs were evaluated to detect phenol, catechol, bisphenol A, and hydroquinone via square wave voltammetry. The sensor showed a detection limit of 1 µM for all the species and was stable and reproducible. It was also successfully used in the simultaneous detection of these species in spiked tap water samples [37]. GCE was modified with AuNPs, L-cysteine, and ZnS/NiS@ZnS quantum dots. It was observed by differential pulse voltammetry and cyclic voltammetry that the modified electrode could be used for concurrent determination of HQ and CA, respectively. The detection limit noted was 24 and 71 nM for HQ and CA and successfully applied for real water samples [38]. Ma et al. developed a virtually monodisperse Au-graphene nanocomposite in an aqueous dimethylformamide solution. GCE modification with the developed nanocomposite showed satisfactory electrocatalytic activity and electronic transport properties because of enhanced surface area. The prepared nanocomposite was also employed to determine CA and HQ with an LOD of 0.15 and 0.2 µM, respectively [39]. Carbon nanocages (CNCs) with AuNPs were prepared by mixing their respective solutions. The AuNPs@ CNC decorated electrodes were used for the sensitive detection of CA and HQ via CV and DPV methods. A satisfactory LOD of 0.0986 µM and 0.0254 µM was observed for CA and HQ, while the developed method was also adequately applied for practical application [40]. Chitosan/N, S co–doped MWCNTs composite with AuNPs were used to modify GCE in preparing the electrochemical-based sensor for CA and nitrate detection (Figure 9.1). The prepared sensor displayed a satisfactory LOD for nitrate and CA, namely, 0.2 µM, with a linear relationship of 1 - 7000 µM and 1 - 5000 µM, respectively. The prepared sensor showed noteworthy selectivity, stability, and reproducibility. The modified electrode was also applied for nitrite and CA determination in tap water and food samples [41]. Table 9.1 represents the electrodes employed to detect phenolic compounds in water.

EG = Ethyl green, DPASV = differential pulse anodic stripping voltammetry, SWASV = square wave anodic stripping voltammetry, SPCE = screen printed carbon electrode, EIS = electrochemical impedance spectroscopy, AuNSs = gold nanostar, SWSV = square wave stripping voltammetry, SH = L–cysteine, SPGE = screen printed gold electrode, ASV = anodic stripping voltammetry.

9.3 DETECTION OF HEAVY METAL (HMS)

Due to their toxicity and detrimental effects, contamination by heavy metal ions has posed a severe threat to animals, plants, and human health [51, 52]. If they are exposed to the environment, they can contaminate water bodies, enter the food chain, and damage animals and humans. They can also have lethal effects on soil and plants accumulating in the soil, causing fertility complications and, result in

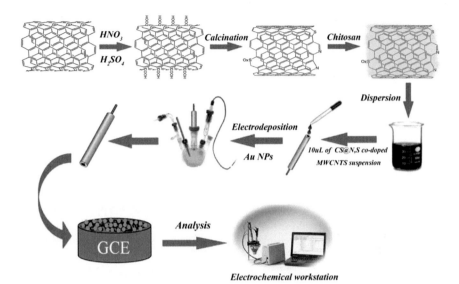

FIGURE 9.1 Preparation of chitosan@N, S, Co–doped MCNTs for the electrochemical sensing of nitrate and CC (Reprinted with permission from ref 41. Copyright 2017. American Chemical Society).

TABLE 9.1
Electrodes Employed for the Detection of Phenolic Compounds in Water

Electrode	Electro-chemical method	Analyte	Detection limit	Linear range	Ref
CoONPs/PBI/ MWCNT/GCE	CV	Phosphates	0.1 nM	0.1 - 100 nM	42
SSE/Cu/SSE/CuPb	CV	Nitrate	5 µM	5 - 1000 µM	43
Co	–	Phosphate	10^{-4} mol/L	10^{-1} - 10^{-4} mol/L	44
Co/Graphene	CV	Phosphate	–	10^{-6} - 10^{-2} M	45
Cu/Ag/PGE	CA	Nitrate	0.034 µM	1 - 10 µM	46
IIP–Cu–NPs/ PANI	CV, LSV	Nitrate	LSV - 5 µM	1 µM - 0.1 M	47
AgNPs/CNTs/GCE	CV	o–cresol, phenol	0.01 µM	10 - 200 µM, 10 - 160 µM	48
TACoPc/PANI/ AgNPs/GCE	CV	Hydroquinone Catechol	0.60 µM, 0.40 µM	10 - 100 µM	49
AuNPs/MWCNTs/ GCE	CV and DPV	Hydroquinone catechol	1 µM	4.0-150.00 µM	50

plant death [53, 54]. Various methods have been taken into consideration, including nanomaterials to determine HMs.

9.3.1 Carbon Nanotubes

Carbon nanotubes (CNTs) can successfully detect heavy metals because of their features such as chemical stability, mechanical strength, fast electron transfer rate, and enhanced surface area [55, 56]. A chitosan/CNT film modified GCE was demonstrated for Cu(II), Cd(II), Pb(II) at pH 5 in an acetic acid buffer via a square wave anodic stripping voltammetry technique. The properties such as enhanced electron transfer rate, high surface area, and good conductivity of CNTS create highly selective and sensitive electrodes. The LOD of the method reported was 0.1 ppm for Cu(II), 0.8 ppm for Cd(II), and 0.6 ppm for Pb (II) [57]. Novel hybrid nanocomposites comprising 1D MWCNTs and 2D GO sheets were prepared. The as-prepared nanocomposite possesses excellent solubility in water due to the high hydrophilicity of GO constituents. It was demonstrated that the nanocomposites could be utilized to sense Pb(II) and Cd(II) in terms of anodic stripping voltammetry. The detection limit noted was 0.2 ppm and 0.1 ppm for Pb(II) and Cd(II) with a deposition time of 180s with a linear range from 0.5 to 30 ppm [58]. A simple, economic modification of MWCNTs with β–cyclodextrin via the covalent and physical method through Steglich esterification for the sensing of Pb(II) was reported. The chemically modified MWCNTs–β–cyclodextrin based electrode displayed satisfactory reliability with LOD of 2.3 ppb, whereas the physically modified electrode depicts a LOD of 0.9 ppb with poor reliability. The prepared sensor showed high reproducibility and reusability of about six times [59]. A series of MWCNTs modified with antimony oxide paste electrodes were employed to detect Cd (II) and Pb(II) ions via ASV. To prepare paste electrodes, a composite of MWCNTs was used and then modified with various concentrations of antimony oxide to enhance the sensing capability. Linear curves of 80 - 150 ppb for Cd(II) and 5 - 35 ppb for Pb(II) were observed with the analytical sensitivity of the metal ion mixtures of 1.932 and 2.694μA L/μg, respectively [60].

9.3.2 Carbon Nanotubes/Gold Nanoparticles

Due to its enhanced water permeability and film-forming capability, chitosan is also used in electrodes or in combination with other materials such as CNTs and AuNPs, iron oxides, and bismuth nanoparticles [61 - 63] for enhanced metal detection capabilities. AuNPs possess characteristics such as enhanced surface area, catalytic properties, and electronic conductivities when used in electrodes [64]. An electrochemical sensor based on screen-printed carbon electrode modified AuNPs/polyaniline MWCNTs composites for the simultaneous determination of metal ions was reported. The prepared sensor displayed superior stability, selectivity, repeatability, and reproducibility with LODs of 0.017 ppm, 0.037 ppm, 0.39 ppm for Cu(II), Pb(II), and Zn(II) with a wide linear range from 1 to 180 ppm

[65]. Modification of glassy carbon electrodes by gold and bismuth bimetallic NPs decorated L–cysteine @GO nanocomposites was reported to detect Fe(II) in sea and lake water samples. It was noticed that the intensity of the current peak increases as a function of Fe(II) concentration in the range of 0.2 - 50 µM with a detection limit of 0.07 µM [66]. Crosslinked CS–CNTs thin-film electrodes for the electrochemical detection of heavy metal ions were developed. The LODs observed for Pb(II), Cd (II) and Cu(II) were 0.6 ppm, 0.8 ppm, and 0.1 ppm, respectively. It was observed that for concurrent determination, the sensitivities of Cd (II) and Pb (II) were enhanced in a Cd(II)–Pb(II) mixture system; however, those of Pb(II) and Cu(II) were weakened in a Pb(II)-Cu(II) mixture system [67].

9.3.3 Bismuth Nanoparticles

Bismuth (Bi) has emerged as an exceptional alternate substitute to Hg in Hg-based electrodes for HMs investigation since it is an environmentally friendly material [68, 69]. Electrodes based on bismuth have extremely reproducible and well-defined responses, good H_2 evolution, and high signal and background characteristics [70]. A simple, versatile in situ strategy for the preparation of reduced graphene oxide/bismuth nanocomposites by the modified Hummers method was reported. The BiNPs were uniformly attached over the surfaces of the graphene nanosheets, resulting in good dispersion in solvents. The prepared rGO/Bi nanocomposite has been utilized as the electrode material to detect HMs in water. The sigma LODs at various deposition potentials for Zn(II), Cd(II), Pb(II), and Cu(II) were reported as 2.8 ppm, 17 ppm, 0.55 ppm, and 26 ppm, respectively [71]. Fabrication of Bi-based screen-printed electrodes for the detection of Pb(II) and Cd(II) was reported by Niu and co–workers (Figure 9.2). The fabrication process included Bi–C material synthesis in bulk, milling process optimization, and then

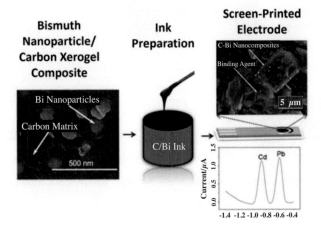

FIGURE 9.2 Scheme for the fabrication of Bi/C SPE sensor (Reprinted with permission from ref 10. Copyright 2021. Electrophoresis, Wiley).

preparing to obtain reproducible and stable electrochemical devices. Calibration curves from voltammograms after 300s gave signals for 10 ppb of Cd(II), and Pb(II), respectively. The peak current increased linearly as a function of metal ions from 10 to 100 ppb [72]. BiNPs were synthesized and loaded on cobalt ferrite for the determination of HMs. The prepared BiNPs@CoFe$_2$O$_4$ nanocomposite was used in modified GCE for the electrochemical determination of HMs ions. The prepared sensor could determine Cd(II) and Pb(II) ions with an LOD of 8.2 and 7.3 nM. The sensor was also convenient for real water samples [73], which possessed acceptable stability and reproducibility. A porous graphene electrode modified with Bi was demonstrated for the electrochemical sensing of Cd(II) in aqueous systems. The free-standing and bicontinuous 3D nanoporous graphene with enlarged surface area and open porosity supported BiNPs. The LOD and limit of quantification of the reported method were calculated to be 0.1 and 0.34 nM. The developed electrode exhibited satisfactory reproducibility, anti-interference capability, and excellent potential for detecting multifold heavy metal ions [74].

9.3.4 Iron Oxides

GCE modifications with iron oxides have proven to be an effective strategy to enhance the electrode response for detecting HMs. For example, magnetite has been employed for chromium determination and various objectives in the water treatment process [75]. A DNA modified Fe$_3$O$_4$@AuNPs, and magnetic GCE-based electrochemical sensing strategy was reported for the rapid and selective detection of Ag(I) and Hg(II) (Figure 9.3). The LOD for the reported method was 3.4 nM for Ag(I) and 1.7 nM for Hg(II), with a signal to noise ratio of 3. Negligible signals

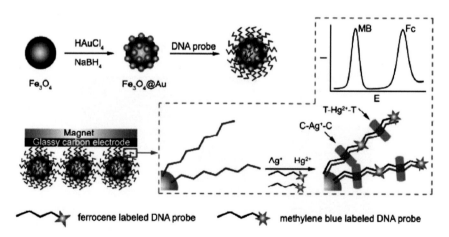

FIGURE 9.3 Schematic representation of DNA modified Fe$_3$O$_4$@AuNPs and for simultaneous detection of Ag(I) and Hg(II) (Reprinted with permission from ref 76. Copyright 2017. American Chemical Society).

Electrochemical Treatment of Pollutants in Water 173

from interfering ions established the satisfactory selectivity of the sensor towards the target ions. The strategy was also used in monitoring the metal ions in various real water samples [76]. Magnetite reduced GO nanocomposites were prepared by a hydrothermal method and were employed to modify the GCE for Cd's sensitive electrochemical determination (II). The LOD with Fe_3O_4–RGO modified GCE towards Cd^{2+} was calculated as 0.056 µM. The developed electrode presented superb stability for Cd(II) electrochemical determination and, therefore, delivered material to enhance the selectivity of electrochemical detection of toxic metal ions [77]. Fe_3O_4@graphene nanocomposite integrated with garlic extract on the GCE surface was used as an active electrochemical probe for Pb(II) sensing. The prepared probe exhibited a LOD of 0.0123 pM and showed significant selectivity, stability, and reproducibility and could also be employed to detect Pb(II) in real water samples [78]. A simple and user-friendly sensor based on core–shell ferroferric oxide@polyaniline NPs was employed for the electrochemical sensing of Cd(II) and Pb(II). With brilliant electronic conductivity, PANI enhanced the electron transport rate and provided good chelation with metal ions. The calculated LODs for Cd(II) and Pb(II) was 0.3 nmol/L and 0.03 nmol/L, respectively [79]. Table 9.2 gives the list of various electrodes used to determine HMs.

9.4 DETECTION OF CHEMICAL OXYGEN DEMAND

Chemical oxygen demand (COD) depicts the level of water pollution and is vital for the valuation of water quality [91]. This method has some disadvantages such as less precision, lower sensitivity, a requirement for a large sample volume, usage of high-cost chemicals, and corrosive H_2SO_4. It also involves a reflux process which is up to 4 hours long, making the method unsuitable for high-level detection [92].

9.4.1 COPPER NANOPARTICLES

Cu is a material used for the electrochemical sensing of COD, as described by various reports. For instance, CNT–polystyrene composite-based electrochemical sensors containing various inorganic electrocatalysts were prepared. The prepared sensors were firstly estimated by glucose as a standard analyte and then by examining COD in water samples from urban water–treatment areas. The expected COD values were also matched with samples from the lab via a $K_2Cr_2O_7$ routine. It was noticed the CuO/AgO-based nanocomposite sensor displayed the best analytical performance and was integrated in a compact flow system showing good agreement with results provided by the $K_2Cr_2O_7$ method [93]. CuNPs were electrodeposited potentiostatically over the surface of GCE for electrochemical assay of COD by using glycine as standard. It was observed by linear sweep voltammetry that CuNPs significantly increase glycine's electrochemical oxidation. The sensor displayed a LOD of 1.7 ppm with a linear range of 12 - 629.3 ppm. It also had a good tolerance level for Cl^- ions (0.35 M). The analytical efficacy was determined by exploring COD recovery and COD assay in various water

TABLE 9.2
Various Electrodes Used for the Detection of Heavy Metal Ions

Electrode	Electrochemical Method	Analyte	LOD	Linear Range	Ref.
ZnO/MWCNTs/GCE	CV	Pb	2.38 µM	4–24 µM	80
PPy/MWCNT	CV	Pb	0.65 ug/L	1–10 µg/L	81
EG/MECNTs/GCE	DPV	Cd	2 nM	2–10 nM	82
MWCNTS/Chitosan/GCE	CV	Cd	0.09 µg/L	–	83
CNTs/Nanohybrids	DPASV	Pb	0.2279 ppb	–	84
		Cu	0.3321 ppb		
GCE/rGO–SH/Au–NPs	DPV	Hg	0.2 µM	1–10 µM	85
AuNPs/MWCNTS/Chitosan	EIS	Cd	0.02 pM	10^{-13}–10^{-4} M	86
AuNSs	SWSV	As(III)	3.6 ppb	0.8–419.4 ppb	87
		Hg(II)	11.1 ppb	33.6–361.5 ppb	
		Pb(II)	20.6 ppb	62.3–215.6 ppb	
Au–BiNPs/SH–GO	CV and EIS	Fe(II)	0.07 µM	0.2–50 µM	66
Chitosan/CNTs	Square–wave anodic stripping voltammetry	Cu(II)	0.1 ppm	0.25–1.24 ppm	67
		Cd(II)	0.8 ppm	1.50–4.44 ppm	
		Pb(II)	0.6 ppm	0.63–3.70 ppm	
PGEs/BiNPs/Nafion	ASV	Pb(II)	31.07 ppm	–	88
		Cd(II)	7.31 µg/L		
SPGE@Nafion	ASV	Pb(II)	3 ppb	20–300 ppb	89
		Cd(II)	4 ppb	50–300 ppb	
BiNCs/AB	–	Pb	1 ng/L	3–4 ppm	90

EG = ethyl green, DPASV = differential pulse anodic stripping voltammetry, SWASV = square wave anodic stripping voltammetry, SPCE = screen–printed carbon electrode, EIS = electrochemical impedance spectroscopy, AuNSs = gold nanostar, SWSV = square wave stripping voltammetry, SH = L–cysteine, SPGE = screen–printed gold electrode, ASV = anodic stripping voltammetry.

samples and verifying the standard $K_2Cr_2O_7$ method [94]. The electrodeposition of nano–Cu was achieved by using a commercially available copper electrical cable and pure Cu disk as a substrate. The nano–Cu modified electrode and bare Cu were demonstrated for the electrochemical assay of COD with glycine as a standard. The oxidation behavior of glycine was noted, and the deposition of nano–Cu enhanced electrochemical oxidation was observed. The prepared nano–Cu sensor possessed a linear range of 2 - 595 µg/mL with a LOD of 1.07 µg/mL with minimal influence of Cl^-(1 M) [95]. Electrochemical reduction of $CuCl_2$ and $Co(NO_3)_2$ resulted in a micro-nano structured Cu-Co fabrication (in situ) over the Au electrode surface. The film prepared at a current of 200 µA in acetate buffer solution (pH 4) depicts high catalytic activity for electrochemical oxidation of glucose. Glucose was used as a standard compound for the evaluation of COD. The sensor has a detection limit of 0.609 μg/L with a 1.92 - 768 μg/L linear concentration range for glucose [96]. A Cu–wire electrode was fabricated to be employed as a sensing electrode for COD detection in water using glycine as a standard. The sensing capability was advanced by CuNPs electrodeposition at optimum preparation conditions. To enhance adherence and stability of deposited CuNPs, the electrode of Cu wire was scratched to increase surface roughness. The proposed sensor had a LOD of 2.6 μg/mL with a linear range of 2 - 595 6 μg/mL and a high tolerance level to Cl^- ions [97].

9.4.2 Graphene Oxide (GO)/Copper Nanoparticles

Graphene has been commonly used as a sensor material because of its special properties. It is generally altered with other constituents to enhance its electrochemical capability [98,99]. A simple, sensitive, and environmentally friendly approach for the electrochemical detection of COD was reported by NiNPs electrodeposited over GCE. Nano–Ni film possessed high catalytic activity and was modified stably over the electrode surface, increasing COD detection sensitivity. Under stable conditions, the LOD of the reported method was as low as 1.1 μg/L along with a linear range of 10 - 1533 μg/L [100]. For enhancing the electrochemical capability, an upright GO screen-printed electrode was prepared with the help of an external magnetic field. NiNPs deposited on the GO electrodes displayed superior oxidation ability for glycine using chronoamperometry. The modified electrode indicated excellent performance for COD analysis with LOD of 0.02 μg/L. The Ni/NPs@ upright GO was also suitable for COD determination in real water samples [101] (Figure 9.4). The GO incorporation as supporting material to NiNPs@nafion GO allowed high catalysts and electrode contact loading,

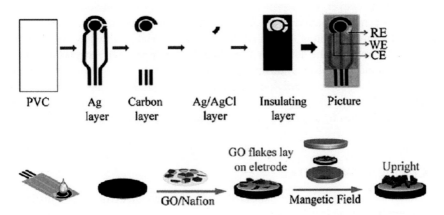

FIGURE 9.4 Diagrammatic representation of NiNPs/upright GO sensor for COD analysis (Reprinted with permission from ref 10. Copyright 2021. Electrophoresis, Wiley).

causing superb electrocatalytic oxidation capability. A flow detection system was developed based on NiNPs@nafion GO screen-printed electrode, which showed a detection limit of 0.05 g/mL and ranged from 0.1 to 400 g/mL. The proposed method was also employed in COD detection for practical applications [102].

9.5 CONCLUSION

This chapter has emphasized recent advancements in preparing electrochemical-based sensors for waste water treatment with a wide range of electrodes. Nanostructured materials in electrochemical sensors have emerged as a substantial investigative tool for sustaining the urge for fast, selective, and on site detection of increasing environmental contaminants. These sensors may simply opt for sensing numerous chemical toxicants and can assist as controllers for upcoming sensing technology. Determining toxicants in water has advanced from expensive and difficult to inexpensive and easy electrochemical approaches. A diversity of nanostructured materials has been established to improve the detection efficacy of the sensors, accomplishing good detection limit, and enhancing abilities such as being convenient for a combination of pollutants and interference resistance. However, to confirm the precision of sensors' capacity in all situations, more investigations or experiments must be performed to identify the interference with other substances. One of the foremost issues that require action in the progress of portable, fast, multi pollutant detection sensors is the huge gap in the research lab benchtop prototypes and their extensive marketable presentations. In ambient conditions, several chemicals possess the ability to combine with others, which increases existing hurdles regarding their identification and toxicology value. As a result, extensive efforts in engineering disciplines and multidisciplinary science will be needed for the efficacious preparation, improvement, and execution of optimized, "market ready" sensors. Nanomaterial-based sensors have been applied in anticipation of a rapid tool for the determination of pollutants. However, many problems must be solved to maintain their success, for instance, mass production of nanomaterials by electrodes for their world-wise usage, the environmental impact of these nanomaterials following use, and so forth. Research into the detection of pollutants should involve the production of cheap nanomaterials to permit mass production, encouraging a combination of nanomaterials to surpass individual weakness, manipulation of the nanomaterials shape to enhance the properties of their materials. Nanomaterials providing new opportunities to determine water pollutants has a great future for this area of study.

ACKNOWLEDGMENTS

The authors are grateful to the Banasthali Vidyapith and the CSIR–Indian Institute of Integrative Medicine, Jammu.

REFERENCES

[1] Zhang, W. Wang, L. Yang, Y. Gaskin, P. and Teng, K. S. "Recent advances on electrochemical sensors for the detection of organic disinfection byproducts in water." *ACS Sensors* 4, no. 5 (2019), 1138–1150. doi:10.1021/acssensors.9b00272.

[2] Perreault, F. de Faria, A. F. and Elimelech, M. "Environmental applications of graphene-based nanomaterials." *Chemical Society Reviews* 44, no. 16 (2015), 5861–5896. doi:10.1039/c5cs00021a.

[3] Su, S. Chen, S. and Fan, C. "Recent advances in two-dimensional nanomaterials-based electrochemical sensors for environmental analysis." *Green Energy and Environment* 3, no. 2 (2018), 97–106. doi:10.1016/j.gee.2017.08.005.

[4] Thangamani, P. Dhanalakshmi, N. Thennarasu, S. and Thinakaran, N. "A novel voltammetric sensor for the simultaneous detection of Cd." *Carbohydrates Polymers* 182, no. 25 (2018), 199–206. doi:10.1016/j.carbpol.2017.11.0172017.

[5] Tanvir, A. and Rangari, S. N. "Graphene oxide-ZnO nanocomposite modified electrode for the detection of phenol." *Analytical Methods* 10, no. 3 (2018), 347–358. doi:10. 1039/c7ay02650a.

[6] Álvarez-Ruiz, R. and Pico, Y. "Analysis of emerging and related pollutants in aquatic biota." *Trends in Environmental Analytical Chemistry* 25, no. 111 (2020), 1–50. doi:10.1016/j.teac.2020. e00082.

[7] Ali, H. Khan, E. and Ilahi, I. "Environmental chemistry and ecotoxicology of hazardous heavy metals: environmental persistence, toxicity, and bioaccumulation." *Journal of Chemistry*, 2019 (2019), 2–19, https://doi.org/10.1155/2019/6730305.

[8] Kapahi, M. and Sachdeva, S. "Bioremediation options for heavy metal pollution". *Journal of Health and Pollution* 9, no. 24 (2019), 1–19. doi: 10.5696/2156-9614-9.24.191203.

[9] Sharma, S. and Bhattacharya, A. Drinking water contamination and treatment techniques. *Applied Water Science* 7, no. 3 (2016), 1043–1067. doi:10.1007/s13201-016-04557.

[10] Potes-Lesoinne, H. A. Ramirez-Alvarez, F. Perez-Gonzalez, V. H. Martinez-Chapa, S. O. and Gallo-Villanueva, R. C. "Nanomaterials for electrochemical detection of pollutants in water: a review." *Electrophoresis* 43, no. 1–2 (2021), 249–262. doi: 10.1002/elps.202100204.

[11] Lingamdinne, L. P. Koduru, J. R. and Karri, R. R. "A comprehensive review of applications of magnetic graphene oxide-based nanocomposites for sustainable water purification." *Journal of Environmental Management* 231 (2019), 622–634. doi:10.1016/j. jenvman.2018.10.063.

[12] Ullah, N. Mansha, M. Khan, I. and Qurashi, A. "Nanomaterial-based optical chemical sensors for the detection of heavy metals in water: recent advances and challenges." *TrAC Trends in Analytical Chemistry* 100 (2018), 155–166. doi:10.1016/j. trac.2018.01.002.

[13] Zaynab, M. Al-Yahyai, R. Ameen, A. Sharif, Y. Ali, L. Fatima, M. AliKhan, K. and Li, S. "Health and environmental effects of heavy metals." *Journal of King Saud University—Science* 34, no. 1 (2022), 1–15. doi: 10.1016/j.jksus.2021.101653.

[14] Molina, J. Cases, F. and Moretto, L. M. "Graphene-based materials for the electrochemical determination of hazardous ions." *Analytical Chimica Acta* 946 (2016), 9–39. doi:10.1016/j.aca.2016.10.019.

[15] Zuo, Y. Xu, J. Zhu, X. Duan, X. Lu, L. and Yu, Y. "Graphene-derived nanomaterials as recognition elements for electrochemical determination of heavy metal ions: a review." *Microchimica Acta* 186, no. 3, 2019, 1–13, doi: 10.1007/s00604019-3248-5.

[16] Yang, Y. Fang, D. Liu, Y. Liu, R. Wang, X. Yu, Y. Zhi, J. "Problems analysis and new fabrication strategies of mediated electrochemical biosensors for waste water toxicity assessment." *Biosensors and Bioelectronics* 108 (2018), 82–88. doi:1016/j.bios.2018.02.049.

[17] Justino, C. I. L. Duarte, A. C. and Rocha-Santos, T. A. P. "Recent progress in biosensors for environmental monitoring: a review." *Sensors* 17, no. 12 (2017), 1–10. doi: 10.3390/s17122918.

[18] Ge, L. Li, S-P. and Lisak, G. "Advanced Sensing Technologies of phenolic compounds for pharmaceutical and biomedical analysis." *Journal of Pharmaceutical and Biomedical Analysis* 179 (2020), 1–13. doi: 10.1016/j.jpba.2019.112913.

[19] Silwana, B. Van Der Horst, C. Iwuoha, E. and Somerset, V. "A brief review on recent developments of electrochemical sensors in environmental application for PGMs." *Journal of Environmental Science and Health—Part A Toxic/Hazardous Substances and Environmental Engineering* 51, no. 14 (2016), 1233–1247. doi: 10.1080/10934529.2016.1212562.

[20] Smith, A. T. LaChance, A. M. Zeng, S. Liu, B. and Sun, L. "Synthesis, properties, and applications of graphene oxide/reduced graphene oxide and their nanocomposites." *Nano Materials Science* 1, no. 1 (2019), 31–47. doi: 10.1016/j.nanoms. 2019.02.004.

[21] Bansod, B. K. Kumar, T. Thakur, R. Rana, S. and Singh, I. "A review on various electrochemical techniques for heavy metal ions detection with different sensing platforms." *Biosensors Bioelectronics* 94 (2017), 443–455. doi: 10.1016/j.bios.2017.03.03.

[22] MA, X. Li, M. Liu, X. Wang, L. Chen, N. Li, J. and Feng, C. "A graphene oxide nanosheet—modified Ti nanocomposite electrode with enhanced electrochemical property and stability for nitrate reduction." *Chemical Engineering Journal* 348 (2018), 171–179. doi: https://doi.org/10.1016/j.cej.2018.04.168.

[23] Liu, F. Liu, K. Li, M. Hu, S. Li, J. Lei, X. and Liu, X. "Fabrication and characterization of a Ni-TNTA bimetallic nanoelectrode to electrochemically remove nitrate from groundwater." *Chemosphere* 223 (2019), 560–568. doi: 10.1016/j.chemosphere.2019.02.028.

[24] Song, Q. Li, M. Wang, L. Ma, X. Liu, F. and Liu, X. "Mechanism and optimization of electrochemical system for simultaneous removal of nitrate and ammonia." *Journal of Hazardous Materials* 363 (2019), 119–126. doi: 10.1016/j.jhazmat.2018.09.046.

[25] Dima, G. E. De Vooys, A. C. A. and Koper, M. "Electrocatalytic reduction of nitrate at low concentration on coinage and transition—metal electrodes in acid solutions." *Journal of Electroanalytical Chemistry* 554, no. 1 (2003) 15–23. doi: 10.1016/S0022–0728(02)01443–2.

[26] Wu, Y. Gao, M. Li, S. Ren, Y. and Qin, G. "Copper wires with seamless 1D nanostructures: preparation and electrochemical sensing performance." *Materials Letters* 211 (2018), 247–249. doi: 10.1016/j.matlet.2017.10.016.

[27] Patella, B. Russo, R. R. O'Riordan, A. Aiello, G. Sunseri, C. and Inguanta, R. "Copper nanowire array as highly selective electrochemical sensor of nitrate ions in water." *Talanta* 221 (2021), 1–12. doi: 10.1016/j.talanta.2020.121643.
[28] Liang, J. Zheng, Y. and Liu, Z. "Nanowire-based Cu electrode as electrochemical sensor for detection of nitrate in water." *Sensors and Actuators B Chemical* 232, 121643-121649 (2016). doi:10.1016/j.snb.2016.03.145.
[29] Ozer, T. O. Özdurak, B. and Doğan, H. Ö. "Electrochemical reduction of nitrate on graphene modified copper electrodes in alkaline media." Journal of Electroanalytical Chemistry, 699 (2013), 1–5. doi: 10.1016/j.jelechem.2013.04.001
[30] Xu, F. Wang, P. Bian, S. Wei, Y. Kong, D. and Wang, H. "A co-nanoparticles modified electrode for on-site and rapid phosphate detection in hydroponic solutions." *Sensors*, 21, no. 1 (2021), 1–22. doi: 10.3390/s21010299.
[31] Pang, H. Cai, W. Ce, S. and Zhang, Y. "Preparation of a cobalt–Fe^{2+}-based phosphate sensor using an annealing process and its electrochemical performance." *Electrochemistry Communications* 124, no. 2 (2021), 1–8. doi: 10.1016/j.elecom.2021.106933.
[32] Wang, X. Ma, X. Church, J. Jung, S. Son, Y. Lee, W. H. and Cho, H. J. "ZnO nanoflakes as a template for *in-situ* electrodeposition of nanostructured cobalt electrodes as amperometric phosphate sensors." *Material Letters* 192 (2017), 107–110. doi: 10.1016/j.matlet.2016.12.047.
[33] Zhu, L. Zhou, X. and Shi, H. "A potentiometric cobalt-based phosphate sensor based on screen-printing technology." *Frontiers of Environmental Science and Engineering* 8, no. 6 (2013), 945–951. doi: 10.1007/s11783-013-0615.
[34] Jiang, Z-J. Liu, C-Y. and Sun, L-W. "Catalytic properties of silver nanoparticles supported on silica spheres." *Journal of Physical Chemistry B*, 109, no. 5 (2005), 1730–1735. doi: 10.1021/jp046032g.
[35] Daniel, M-C. and Astruc, D. "Gold Nanoparticles: assembly, supramolecular chemistry, quantum-size-related properties, and applications toward biology, catalysis, and nanotechnology." *Chemical Reviews* 104, no. 1 (2004), 293–346. doi: 10.1021/cr030698+.
[36] Ortiz-Castillo, J. E. Gallo-Villanueva, R. C. Madou, M. J. and Perez-Gonzalez, V. H. "Anisotropic gold nanoparticles: a survey of recent synthetic methodologies." *Coordination Chemistry Reviews* 425 (2022), 1–23. doi: 10.1016/j.ccr.2020.213489.
[37] Goulart, L. A. Gonçalves, R. Correa, A. A. Pereira, E. C. and Mascaro, L. H. "Synergic effect of silver nanoparticles and carbon nanotubes on the simultaneous voltammetric determination of hydroquinone, catechol, bisphenol A and phenol." *Microchimica Acta* 185, no. 12 (2018), 1–13.
[38] Wang, Y. Qu, J. Li, S. Dong, Y. and Qu, J. "Simultaneous determination of hydroquinone and catechol using a glassy carbon electrode modified with gold nanoparticles, ZnS/NiS@ZnS quantum dots and L-cysteine." *Microchimica Acta* 182 (2016), 2277–2283.
[39] Ma, X. Liu, Z. Qiu, C. and Chen, T. "Simultaneous determination of hydroquinone and catechol based on glassy carbon electrode modified with gold-graphene nanocomposite." *Microchimica Acta* 180, no. 5–6 (2013). doi: 10.1007/s00604-013-0949-z.
[40] Huang, Y. H. Chen, J. H. Ling, L. J. Su, Z. B. Sun, X. Hu, S. R. Weng, W. Huang, Y. Wu W. B. and He, Y. S. "Simultaneous electrochemical detection of catechol

and hydroquinone based on gold nanoparticles@carbon nanocages modified electrode." *Analyst* 140 (2015), 7939–7947. doi: 10.1039/C5AN01738F.

[41] Rao, H. Liu, Y. Zhong, J. Zhang, Z. Zhao, X. Liu, X. Jiang, Y. Zou, P. Wang, X. and Wang, Y. "Gold nanoparticle/chitosan@N,S Co-doped multiwalled carbon nanotubes sensor: fabrication, characterization, and electrochemical detection of catechol and nitrite." *ACS Sustainable and Chemical Engineering* 5, no. 11 (2017), 010926–10939. doi: 10.1021/acssuschemeng.7b02840.

[42] Sari, S. R. Tsushida, M. Sato, T. and Tominaga, M. "Highly sensitive detection of phosphate using well-ordered crystalline cobalt oxide nanoparticles supported by multi-walled carbon nanotubes." *Material Advances* 3, no. 4 (2022), 1–8. doi: 10.1039/D1MA01097B

[43] Shao, Y-A. Chen, Y-T. and Chen, P-Y. "Cu and CuPb electrodes prepared via potentiostatic electrodeposition from metal oxides in hydrophobic protic amide-type ionic liquid/water mixture under ambient air for nonenzymatic nitrate reduction." *Electrochimica Acta* 313 (2019), 488–496. doi: 10.1016/j.electacta.2019.05.057

[44] Xu, F. Wang, P. Bian, S. Wei, Y. Kong, D. and Wang, H. "Co-nanoparticles modified electrode for on-site and rapid phosphate detection in hydroponic solutions." *Sensors* 21, no. 1 (2021). doi: 10.3390/s21010299.

[45] Kazem, K. Payam, H. and Hamid, A. S. "Cobalt-graphene nanocomposite electrode for phosphate sensing." *Analytical and Bioanalytical Electrochemistry* 9, no. 5 (2017), 521–534.

[46] Naveen, B. and Kumar, S. "Templated bimetallic copper–silver nanostructures on pencil graphite for amperometric detection of nitrate for aquatic monitoring." *Journal of Electroanalytical Chemistry* 856 (2020), 1–10. doi: 10.1016/j.jelechem.2019.113660.

[47] Essousi, H. Barhoumi, H. Bibani, M. Ktari, N. Wendler, F. Al-Hamry, A. and Kanoun, O. "Ion-imprinted electrochemical sensor based on copper nanoparticles-polyaniline matrix for Nitrate Detection. *Hindawi Journal of Sensors* 2019, 1–14. doi: 10.1155/2019/4257125.

[48] Zhu, Y. and Yang, L. "Synthesis of Ag nanoparticles decorated carbon nanotubes as an electrochemical sensor for determination of phenolic compounds in shale gas waste water." *International Journal of Electrochemical Science* 16 (2021), 1–10. doi: 10.20964/2021.07.10.

[49] Sudhakara, S. M. Devendrachari, M. C. Kotresh, H. M. N. and Khan, F. "Silver nanoparticles decorated phthalocyanine doped polyaniline for the simultaneous electrochemical detection of hydroquinone and catechol." *Journal of Electroanalytical Chemistry* 884 (2021), 1–14. doi: 10.1016/j.jelechem.2021.115071.

[50] Piña, S. Candia-Onfray, C. Hassan, N. Jara-Ulloa, P. Contreras, D. and Salazar, R. "Glassy carbon electrode modified with C/Au nanostructured materials for simultaneous determination of hydroquinone and catechol in water matrices." *Chemosensors* 9, no. 5 (2021), 1–17. doi: 10.3390/chemosensors9050088.

[51] Needleman, H. "Lead poisoning." *Annual Review of Medicine* 55 (2004), 209–222. doi: 10.1146/annurev.med.55.091902.103653.

[52] Mishra, S. Bharagava, R. N. More, N. Yadav, A. Zainith, S. Mani, S. and Chowdhary, P. "Heavy metal contamination: an alarming threat to environment

and human health." *Springer, Singapore* (2019), 103–125.doi: 10.1007/978-981-10-7284-0_5.

[53] Somashekaraiah, B. V. Padmaja, K. and Prasad, A. R. K. "Phytotoxicity of cadmium ions on germinating seedlings of mung bean (*Phaseolus vulgaris*): involvement of lipid peroxides in chlorophyll degradation." *Physiologia Plantarum* 85, no. 1 (1995), 85–89. doi:10.1111/j.1399-3054.1992.tb05267.x.

[54] Guo, J. Dai, X. Xu, W. and Ma, M. "Overexpressing *GSH1* and *AsPCS1* simultaneously increases the tolerance and accumulation of cadmium and arsenic in *Arabidopsis thaliana.*" *Chemosphere* 72 (2008), 1020–1026. doi: 10.1016/j.chemosphere.2008.04.018.

[55] Gooding, J. J. Wibowo, R. Liu, J. Yang, W. Losic, D. Orbons, S. Mearns, F. J. Shapter, J. G. and Hibbert, D. B. "Protein electrochemistry using aligned carbon nanotube arrays." *Journal of American Chemical Society* 125, no. 30 (2003), 9006–9007.doi: 10.1021/ja035722f.

[56] Hwang, G. H. Han, W. K. Park, J. S. and Kang, S. G. "Determination of trace metals by anodic stripping voltammetry using a bismuth-modified carbon nanotube electrode." *Talanta* 76, no. 2 (2015), 301–308. doi: 10.1016/j.talanta.2008.02.039.

[57] Huanga, H. Zhua, W. Gaoa, X. Liub, X. and Maa, H. "Synthesis of a novel electrode material containing phytic acid–polyaniline nanofibers for simultaneous determination of cadmium and lead ions." *Analytical Chimica Acta* 852 (2014), 45–54. doi: 10.1016/j.aca.2014.09.010.

[58] Wu, K-H. Lo, H-M. Wang, J-C. Yu, S-Y. and Yang, B-D. "Electrochemical detection of heavy metal pollutant using crosslinked chitosan/carbon nanotubes thin film electrodes." *Materials Express*, 7, no. 1 (2017), 15–24. doi: https://doi.org/10.1166/mex.2017.1351.

[59] Alam, A. U. Howlader, M. M. R. Hu, N-X. and Deen, M. J. "Electrochemical sensing of lead in drinking water using β-cyclodextrin-modified MWCNTs." *Sensors and Actuator B: Chemical* 296 (2019), 1–13. doi: 0.1016/j.snb.2019.126632.

[60] Hai, T. L. Hung, L. C. Phuong, T. T. B. Ha, B. T. T. Nguyen, B. S. Hai, T. D. and Nguyen, V. H. "Multiwall carbon nanotube modified by antimony oxide (Sb_2O_3/MWCNTs) paste electrode for the simultaneous electrochemical detection of cadmium and lead ions." *Microchemical Journal* 153, 2019, 1–14. doi: 10.1016/j.microc.2019.104456.

[61] Sugunan, A. Thanachayanont, C. Dutta, J. and Hilborn, J. "Heavy-metal ion sensors using chitosan-capped gold nanoparticles." *Science and Technology of Advanced Materials* 6, no. 3–4 (2005), 335–340. doi: 10.1016/j.stam.2005.03.007.

[62] Ahmed, R. A. and Fekry, A. M. "Preparation and characterization of a nanoparticles modified chitosan sensor and its application for the determination of heavy metals from different aqueous media." *International Journal of Electrochemical Science* 8, no. 5 (2013), 6692–6708.

[63] Hwang, J-H. Pathak, P. Wang, X. Rodriguez, K. L. Cho, H. J. and Lee, W. H. "A novel bismuth-chitosan nanocomposite sensor for simultaneous detection of Pb(II), Cd(II) and Zn(II) in waste water." *Micromachines* 10, no. 8 (2019), 1–7. doi: 10.3390/mi10080511.

[64] Dai, X. Nekrassova, O. Hyde, M. E. and Compton, R. G. "Anodic stripping voltammetry of Arsenic(III) using gold nanoparticle-modified electrodes." *Analytical Chemistry* 76, no. 19 (2004), 5924–5929. doi:10.1021/ac049232x.

[65] Shao, Y. Dong, Y. Bin, L. Fan, L. Wang, L. Yuan, X. Li, D. Liu, X. and Zhao, S. "Application of gold nanoparticles/polyaniline-multi-walled carbon nanotubes modified screen-printed carbon electrode for electrochemical sensing of zinc, lead, and copper." *Microchemical Journal* 170, 106726-106732 (2021). doi:10.1016/j.microc.2021.106726.

[66] Zhou, N. Li, J. Wang, S. Zhuang, X. Ni, S. Luan, F. Wu, X. and Yu, S. "An electrochemical sensor based on gold and bismuth bimetallic nanoparticles decorated l-cysteine functionalized graphene oxide nanocomposites for sensitive detection of iron ions in water samples." *Nanomaterials* 11 (2021), 1–11. doi: 10.3390/nano11092386.

[67] Wu, K-H. Lo, H-M. Wang, J-C. Yu, S-Y. and Yan, B-D. "Electrochemical detection of heavy metal pollutant using cross-linked chitosan/carbon nanotubes thin film electrodes." 7, no. 1 (2017), 15–24, doi: 10.1166/mex.2017.1351.

[68] Hwang, G. H. Han, W. K. Park, J. S. and Kang, S. G. "Determination of trace metals by anodic stripping voltammetry using a bismuth-modified carbon nanotube electrode". *Talanta* 76, no. 2 (2008), 301–308, doi: 10.1016/j.talanta.2008.02.039.

[69] Lee, G-J. Lee, H. M. Uhm, Y. R. Lee, M. K. and Rhee, C-K. "Square-wave voltammetric determination of thallium using surface modified thick-film graphite electrode with Bi nanopowder." *Electrochemistry Communications* 10, no. 12 (2008), 1920–1923. doi: 10.1016/j.elecom.2008.10.015.

[70] Wang, J. "Stripping analysis at bismuth electrodes: a review." *Electroanalysis* 17, no. 15–16 (2005), 1341–1346. doi: 10.1002/elan.200403270.

[71] Sahoo, P. K. Panigrahy, B. Sahoo, A. K. Li, S. D. and Bahadur, D. "*In situ* synthesis and properties of reduced graphene oxide/Bi nanocomposites: as an electroactive material for analysis of heavy metals." *Biosensors and Bioelectronics* 43 (2013), 293–296, doi: 10.1016/j.bios.2012.12.031.

[72] Niu, P. Fernández-Sánchez, C. Gich, M. Navarro-Hernández, C. Fanjul-Bolado, P. and Roig, A. "Screen-printed electrodes made of a bismuth nanoparticle porous carbon nanocomposite applied to the determination of heavy metal ions." *Microchimica Acta* 183 (2016), 617–623. doi: 10.1007/s00604-015-1684-4.

[73] He, Y. Wang, Z. Ma, L. Zhou, L. Jiang, Y. and Gao, J. "Synthesis of bismuth nanoparticle-loaded cobalt ferrite for electrochemical detection of heavy metal ions." *RSC Advances* 10 (2020), 27697–27705. doi: 10.1039/D0RA02522D.

[74] Huang, L. Ito, Y. Fujita, T. Ge, X. Zhang, L. and Zeng, H. "Bismuth/porous graphene heterostructures for ultrasensitive detection of Cd(II)." *Materials* 13, no. 22 (2020), 1–13. doi: 10.3390/ma13225102.

[75] Prakash, A. Chandra, S. and Bahadur, D. "Structural, magnetic, and textural properties of iron oxide-reduced graphene oxide hybrids and their use for the electrochemical detection of chromium." *Carbon* 50 (2012), 114209–114219. doi: 10.1016/j.carbon.2012.05.002.

[76] Miao, P. Tang, Y. and Wang, L. "DNA modified $Fe3O_4$@Au magnetic nanoparticles as selective probes for simultaneous detection of heavy metal ions." *ACS Applied Materials Interfaces* 9, no. 4 (2017), 3940–3947. doi: 110.1021/acsami.6b14247.

[77] Sun, Y-F. Chen, W-K. Li, W-J. Jiang, T-J. Liu, J-H. and Liu, Z-G. "Selective detection toward Cd^{2+} using Fe_3O_4/RGO nanoparticle modified glassy carbon electrode." *Journal of Electroanalytical Chemistry* 714–715 (2014), 97–102. doi:10.1016/j.jelechem.2013.12.030.

[78] He, B. Shen, X-F. Nie, J. Wang, X-L. Liu, F-M. Yin, W. Hou, C-J. Huo, D-Q. and Fa, H-B. "Electrochemical sensor using graphene/Fe_3O_4 nanosheets functionalized with garlic extract for the detection of lead ion." *Journal of Solid-State Electrochem*istry 11 (2018), 3515–3525. doi: 10.1007/s10008-018-4041-9.

[79] Kong, Y. Wu, T. Wu, D. Zhang, Y. Wang, Y. Du, B. and Wei, Q. "An electrochemical sensor based on Fe_3O_4@PANI nanocomposites for sensitive detection of Pb^{2+} and Cd^{2+}." *Analytical Methods* 10 (2018), 14784–14792. doi: 10.1039/C8AY01245H.

[80] Malik, L. A. M. Pandith, A. H. Bashir, A. and Qureashi, A. "Zinc oxide-decorated multiwalled carbon nanotubes: a selective electrochemical sensor for the detection of Pb(II) ion in aqueous media" *Journal of Material Science: Materials in Electronics* (2022). doi: 10.1007/s10854-022-07793-x.

[81] Zhu, X. Tong, J. Bian, C. Gao, C. and Xia, S. "The polypyrrole/multiwalled carbon nanotube modified au microelectrode for sensitive electrochemical detection of trace levels of Pb^{2+}." *Micromachanics* (2017), no. 86, 1–9, doi: 10.3390/mi8030086.

[82] Sreekanth, A. P. Alodhayb, A. Assaifan, A. K. Alzahrani, K. E. Muthuramamoorthy, M. Alkhammash, H. I. Pandiaraj, S. Alswieleh, A. M. Van Le, Q. Mangaiyarkaras, R. Grace, A. N. and Raghavan, V. "Multi-walled carbon nanotube-based nanobiosensor for the detection of cadmium in water." *Environmental Research* 197 (2021), 1–11. doi: 10.1016/j.envres.2021.111148.

[83] Solís, C. and Galicia, M. "High Performance of MWCNTs—chitosan modified glassy carbon electrode for voltammetric trace analysis of Cd(II)." *International Journal of Electrochemical Science* 15 (2020), 6815–6828. doi: 10.20964/2020.07.56.

[84] Silva, A. C. O. de Oliveira, L. C. F. Delfino, A. V. Meneghetti, M. R. and de Abreu, F. C. "Electrochemical study of carbon nanotubes/nanohybrids for determination of metal species Cu^{2+} and Pb^{2+} in water samples." *Journal of Analytical Methods in Chemistry* 2016 (2016), 1–12. doi: 10.1155/2016/9802738.

[85] Devi, N. R. Sasidharan, M. and Sundramoorthy, A. K. "Gold nanoparticles-thiol-functionalized reduced graphene oxide coated electrochemical sensor system for selective detection of mercury ion." *Journal of the Electrochemical Society* 165, no. 8 (2018), B3046–B3053. doi: 10.1149/2.0081808jes.

[86] Rebai, S. Teniou, A. Catanante, G. Benounis, M. Marty, J-L. and Rhouati, A. "Fabrication of AuNPs/MWCNTS/Chitosan nanocomposite for the electrochemical aptasensing of cadmium in water." *Sensors* 22, no. 1 (2021), 1–10. doi: 10.3390/s22010105.

[87] Dutta, S. Strack, G. and Kurup, P. "Gold nanostar electrodes for heavy metal detection." *Sensors and Actuators B: Chemical* 281 (2018), 383–391. doi: 10.1016/j.snb.2018.10.111.

[88] Palisoc, S. Gonzales, A. J. Pardilla, A. Racines, L. and Natividad, M. "Electrochemical detection of lead and cadmium in UHT-processed milk using bismuth nanoparticles/Nafion®-modified pencil graphite electrode." *Sensing and Bio-sensing Research* 23 (2019), 1–12. doi: 10.1016/j.sbsr.2019.100268.

[89] Albalaw, I. Hogan, A. Alatawi, H. and Moore, E. "A sensitive electrochemical analysis for cadmium and lead based on Nafion-Bismuth film in a water sample." *Sensing and Bio-sensing Research* 34 (2021), 1–19. doi: 10.1016/j.sbsr.2021.100454.

[90] Zou J. Yu, Q. Gao, Y. Chen, S. Huang, X. Hu, D. Liu, S. and Lu, L-M. "Bismuth nanoclusters/porous carbon composite: a facile ratiometric electrochemical sensing platform for Pb^{2+} detection with high sensitivity and selectivity." *ACS Omega* 7, no. 1 (2022), 1132–1138. doi: 10.1021/acsomega.1c05713.

[91] Li, J. Luo, G. He, L. J. Xu, J. and Lyu, J. "Analytical approaches for determining chemical oxygen demand in water bodies: a review." *Critical Reviews in Analytical Chemistry* 48, no. 1 (2018), 47–65. doi: 10.1080/10408347.2017.1370670.

[92] Kabir, H. Zhu, H. Lopez, R. Nicholas, N. W. McIlroy, D. N. Echeverria, E. May, J. and Cheng, I. F. "Electrochemical determination of chemical oxygen demand on functionalized pseudo-graphite electrode." *Journal of Electroanalytical Chemistry* 851 (2019), 1–15. doi: 10.1016/j.jelechem.2019.113448.

[93] Gutiérrez-Capitán, M. Baldi, A. Gómez, R. García, V. Jiménez-Jorquera, C. and Fernández-Sánchez, C. "Electrochemical nanocomposite-derived sensor for the analysis of chemical oxygen demand in urban waste waters." *Analytical Chemistry* 87, no. 4 (2015), 2152–2160. doi: 10.1021/ac503329a.

[94] Badr, I. H. A. Hassan, H. H. Hamed, E. and Abdel-Aziz, A. M. "Sensitive and green method for determination of chemical oxygen demand using a nano-copper based electrochemical sensor." *Electroanalysis* 29, no. 10 (2017), 2401–2409. doi: 10.1002/elan.201700219.

[95] Hassan, H. Badra, I. Hesham T. Abdel-Fatah, M. Elfeky, E. M. S. and Abdel-Aziz, A. M. "Low cost chemical oxygen demand sensor based on electrodeposited nano-copper film." *Arabian Journal of Chemistry* 11, no. 2 (2018), 171–180. doi: 10.1016/j.arabjc.2015.07.001.

[96] Wang, J. Yao, N. Li, M. Hu, J. Chen, J. Hao, Q. Wu, K. and Zhou, Y. "Electrochemical tuning of the activity and structure of a copper–cobalt micro–nano film on a gold electrode, and its application to the determination of glucose and of chemical oxygen demand." *Microchimica Acta* 182 (2015), 515–522. doi: 10.1007/s00604-014-1353.

[97] Elfeky, E. M. S. Shehata, M. R. Elbashar, Y. H. Barakat, M. H. and El Rouby, W. M. A. "Developing the sensing features of copper electrodes as an environmental friendly detection tool for chemical oxygen demand." *RSC Advances* 12, no. 7 (2022), 4199–4208. doi: 0.1039/D1RA09411D.

[98] Lawal, L A. "Progress in utilization of graphene for electrochemical biosensors." *Biosensors and Bioelectronics* 106 (2018), 149–178. doi: 10.1016/j.bios.2018.01.030.

[99] Zeng, L. Cao, S. Yin, H. Xiong, J. and Lin, D. "*Graphene* oxide—applications and opportunities." *InTech*, London, UK (2018). doi: 10.5772/intechopen.78222.

[100] Jing, T. Zhou, Y. Hao, Q. Zhou, Y. and Mei, S. "A nano–nickel electrochemical sensor for sensitive determination of chemical oxygen demand." *Analytical Methods* 4 (2012), 1155–1159. doi: 10.1039/c2 ay05631c.

[101] Li, X. Lin, D. Lu, K. Chen, X. Yin, S. Li, Y. Zhang, Z. Tang, M. and Chen, G. "Graphene oxide orientated by a magnetic field and application in sensitive detection of chemical oxygen demand." *Analytica Chimica Acta* 1122 (2020), 31–38. doi: 10.1016/j.aca.2020.05.009.

[102] Zhang, B. Huang, L. Tang, M. Hunter, K. W. Feng, Y. Sun, Q. Wang, J. and Che, G. "A nickel nanoparticle/nafion–graphene oxide modified screen-printed electrode for amperometric determination of chemical oxygen demand." *Microchimica Acta*, 185, no. 385 (2018), 1–9. doi:10.1007/s00604-018-2917-0.

10 Nanotoxicology and Challenges of Using Nanomaterials for Water Treatment

Nitin Kumar Sharma[1,2] and
Jyotsna Vishwakarma[3]
[1]Department of Chemical Engineering, Indian Institute of Technology, Kanpur, India
[2]Shri Maneklal M. Patel Institute of Sciences and Research, Kadi Sarva Vishwavidyalaya, Gandhinagar, Gujarat, India
[3]K.B. Pharmacy Institute of Education and Research, Kadi Sarva Vishwavidyalaya, Gandhinagar, Gujarat, India

CONTENTS

10.1	Introduction	186
10.2	Water and Waste water Treatment Using Nanomaterials	186
	10.2.1 Zero-Valent Metal Nanoparticles	187
	10.2.1.1 AgNPs	187
	10.2.1.2 FeNPs	188
	10.2.1.3 ZnNPs	189
	10.2.2 Metal Oxide Nanoparticles	189
	10.2.3 TiO_2 Nanoparticles	189
	10.2.4 ZnO Nanoparticles	190
	10.2.5 Iron Oxide Nanoparticles	190
	10.2.6 Carbon Nanotubes	191
	10.2.7 Nanocomposites	192
	10.2.8 Polymer-Based Nanoadsorbents	192
10.3	Nanomaterial Toxicity	192
10.4	Nanomaterial Alteration Hazard in Water	193
	10.4.1 Water Pollution	193
10.5	Conclusion	193
Acknowledgements		194
References		194

DOI: 10.1201/9781003252931-10

10.1 INTRODUCTION

Nanotoxicology has been increasing attention from researchers into the manufacturing nanomaterials. Many dangerous exposure situations have been experienced by workers working in nanotechnology research and developments. NMs may have substantial, unknown, dangerous properties that can pose risks for researchers, in transportation, in storage, in waste facilities, and when any NM disasters occur. Thus, nanotoxicology deals with the study of the toxicity of NMs since it is critical to understand the toxicity of NMs before using them for various applications. Nowadays, the greatest challenge faced in nanotoxicology is recognizing and estimating the damaging effects of various engineered NMs because of their different physiochemical properties [1]. For different reasons, it is problematic to find the exact hazard denominations of NPs. For instance, we are not specific about which physicochemical property of the nanoparticles is manipulating the toxicity. Previously, the NMs have been under vigorous development and used effectively in multiple applications, such as catalysts [2], in medical devices and medicines [3], in sensing [4], and in the biological field [5]. Their application in the treatment of waste water has received wide attention around the globe. NMs have very interesting adsorption characteristics and reactivity because of their small sizes, high surface areas, and active mobility in solution [6]. NMs are generally used to eliminate toxic heavy metals [7], organic pollutants [8], inorganic anions [9], and bacteria [10]. It is also very well reported that NMs show a great value for applications in water and waste water handling. For waste water treatment, zero-valent metal nanoparticles, metal oxide nanoparticles, CNTs, and nanocomposites are extensively studied.

10.2 WATER AND WASTE WATER TREATMENT USING NANOMATERIALS

Many NMs are known for their role in water and waste water purification stages. Microorganisms, organic toxicants, metal ions, and the like, can be removed from waste water by selecting various and appropriate NMs (Figure 10.1).

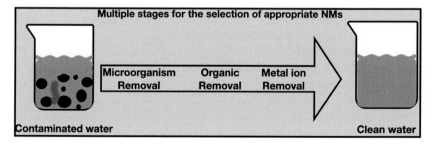

FIGURE 10.1 Various stages for water purification by using an appropriate NMs.

A few of the NMs that can be used as NMs for waste water treatment are discussed here.

10.2.1 ZERO-VALENT METAL NANOPARTICLES

Many zero-valent NPs are known for water purification. A few of them are studied for their wider utilization in removing contaminants (Figure 10.2). The most effective NPs such as AgNPs, FeNPs, and ZnNPs are used, based on their extraordinary properties.

10.2.1.1 AgNPs

The toxic nature of AgNPs against a wide range of microorganisms makes them antibacterial agents, including against fungi or viruses [11,12]. Due to their strong antimicrobial activity, it was also reported that AgNPs are widely studied for the disinfection or purification of water. The AgNPs have conveyed their ability to adhere on the bacterial cell wall and consequently to go through it, resulting in organizational modifications of the cell membrane including an increase in its permeability [13]. Moreover, when AgNPs contact bacteria, many free radicals are generated. Their capability for the destruction of the cell membrane is known to create cell death [14].

Additionally, as DNA contains abundant elements S and P, the AgNPs can destroy it. The death of cells is due to the dissolution of the AgNPs, which will release Ag^+ ions, which can interact with thiol groups of numerous enzymes, disable them, and interrupt systematic cell functions [15, 16]. Recently, AgNPs have been successfully used in the disinfection of both waste water and fresh water.

The tendency to aggregate and reduce efficacy during long-term use are the factors that reduce the direct use of AgNPs in various applications [17]. The functionalized AgNPs and attached materials make them more suitable for water

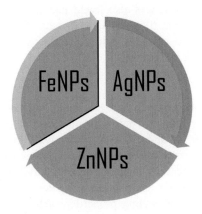

FIGURE 10.2 Various types of NPs used for water and waste water treatment.

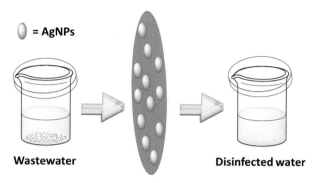

FIGURE 10.3 Waste water treatment by using AgNPs on blotting paper.

purification or disinfection due to their antibacterial activities and cheaper cost [18]. Reduction of $AgNO_3$ produced AgNPs dumped on cellulose fibers of blotting paper, and the sheet expressed great antibacterial activities against E. coli and E. faecalis during sheet filtration (Figure 10.3).

Therefore, selecting paper deposited AgNPs for waste water filtration is a very effective technique for water treatment. Further, AgNPs were prepared through a chemical reduction process and have been included in PES microfiltration. A huge suppression in the activity of microorganisms is observed. It is concluded that the combination of PES-AgNPs membranes exhibited sound antimicrobial activities and held a great antimicrobial tendency for use in waste water management [19].

The AgNPs on ceramic material membranes have received attention globally [20]. For example, AgNPs modified with ceramic filters, and clay trigger the removal of E. coli from water. The porosity is also a noticeable parameter that increases the removal of microorganisms if it is high and decreases the removal if the porosity is low [21]. It was also found that colloidal AgNPs mixed with modified cylindrical ceramic filters enhanced their efficiency, which can now remove 97.8% and 100% E. coli [22]. Additional extensive studies on AgNPs will increase their waste water applications.

10.2.1.2 FeNPs

Previously, Fe, Zn, Al, and Ni and other zero-valent NPs have drawn wide research interest in waste water treatment. Their standard reduction potentials (E^0/V) of Fe, Zn and Ni are -0.440, -0.762 and -0.236 respectively [23].

Lower -ve reduction potential of Ni has a lower standard reduction potential which indicates a lesser reducing character than Fe. Along with a rational standard reduction potential, Zn or Fe has outstanding potential to work as a reducing agent. Fe shows various outstanding properties such as low cost, precipitation, great absorption, and lower reduction ability. Hence, FeNPs have been broadly deliberated NPs. This uniqueness makes them suitable candidates for the removal of contaminants.

Although, several studies have focused on zero-valent metal NPs such as Fe and Zn for waste water treatment [24]. Zn is considered an alternative due to its additional -ve standard reduction potential, making it a more robust reductant than Fe.

10.2.1.3 ZnNPs

Apart from FeNPs, Zn has also proved its dominancy as an alternative for fresh water and waste water treatment [24]. With a more negative standard reduction potential, the higher reduction potential of Zn makes it a more suitable reductant. Thus, the degradation rate of ZnNPs might be quicker than that of other zero-valent NPs. The ZnNPs are mostly studied for dehalogenation reactions. It is found that the rate of degradation of CCl_4 with ZnNPs is found to be more rapid than other zero-valent NPs [25]. Nevertheless, even though numerous results have established that reducing pollutants by ZnNPs could be successful, the applications of ZnNPs are restricted in organic compound degradation, particularly CCl_4.

10.2.2 METAL OXIDE NANOPARTICLES

Metal oxide NPs form another class of materials that has proved its potential for waste water treatment. Such NMs have numerous other possible applications in the decontamination of waste water.

10.2.3 TiO_2 NANOPARTICLES

Fujishima and Honda et al. [26] reported water electrochemical photolysis on TiO_2 semiconductor electrodes in 1972. Recently, photochemical degradation was applied successfully in pollutant elimination from water and waste water. The contaminant breaks down by oxidation into an intermediate of lower molecular weight and is then transformed into neutral CO_2, H_2O, and some anions such as PO_4^{-3}, Cl^- and NO_3^-. TiO_2 has been studied extensively over other photocatalysts such as metal oxide and sulfide semiconductors due to its photostability, biochemical stability, and economy [27]. The induction of charge separation in the particle by UV is required in TiO_2 due to the high bandgap. The ROS generated by TiO_2 is responsible for pollutant degradation upon UV irradiation quickly (Figure 10.2). Due to its selectivity, the TiO_2 NPs are appropriate for all types of contaminants, such as organic compounds, chloro derivatives, polycyclic aromatic hydrocarbons, dyes, phenols, pesticides, arsenic, cyanide, and heavy metals [28 - 35]. Their photocatalytic activities have the potential to destroy a variety of microorganisms, fungi, viruses, and algae [36].

On the other hand, the TiO_2 NPs also have several disadvantages, but, nevertheless, superior activity. They always need UV and visible light for irradiation due to their higher bandgap and their photocatalytic properties. Therefore, under UV and visible light, many investigations have taken place to expand the photocatalytic properties of the TiO_2NPs [37]. The doping of Ag in TiO_2NPs is widely studied because it might facilitate visible light excitation [38] and significantly increase the rate of bacterial and viral photocatalytic inactivation [39].

Further, the procedure for producing TiO_2NPs is rather difficult. The recovery of TiO_2NPs from the tested contaminated water is very complicated when they are used in the suspension. Recently, much hard work has been reported to overcome this difficulty. Among them, a coupling of the photocatalyst of TiO_2NPs with membrane technology has the potential to solve the TiO_2NPs recovery problem from waste water, and hence it has widely attracted researchers. Doped magnetic TiO_2 NPs are produced for the recovery of a uniform morphology of the NPs; a technique based on the spinning of a disc reactor in a magnetic field is used to produce magnetic TiO_2 NPs. [40]. It is a continuous production process and thus appropriate for industrial applications [41].

10.2.4 ZnO Nanoparticles

Another of the most efficient candidates in photocatalysis are ZnO NPs because of their wide bandgap, strong oxidizing ability, biocompatibility, eco-friendliness, and many other unique characteristics in fresh water and waste water treatment [42]. The low cost of ZnO NPs is an advantage over TiO_2NPs. Additionally, the absorption of light quanta and solar spectra made by ZnO NPs over other known semiconducting metal oxides makes them more efficient [43]. The limitation of using ZnO NPs is like TiO_2NPs and is limited due to the high bandgap in UV light regions. The doping of metals is the most common technique to improve photodegradation efficiency. Anionic, cationic, rare earth and co-dopants are the common metals that have been tested positively [44].

10.2.5 Iron Oxide Nanoparticles

Due to the simple properties and wider availability of the FeONPs, they are used for heavy metal removal from waste water treatments. Magnetic forms of maghemite (γ-Fe_2O_4), magnetite (Fe_3O_4), and nonmagnetic hematite (α-Fe_2O_3) are frequently used as nano-adsorbents. The main challenge in removing contaminants from waste water is the small size of the nanosorbent. However, simply by just applying the magnetic field, the magnetic maghemite (γ-Fe_2O_4) and magnetite (Fe_3O_4) can also be used successfully as sorbents in heavy metal removal from waste water [45]. To increase adsorption and to avoid a barrier from extra metal ions, the FeONPs are functionalized to refine their adsorption abilities by the addition of a variety of ligands [46] or polymers [47]. The nature of polymeric molecules is to bind with metal ions and act as carriers from tested water [48]. The hematite (α-Fe_2O_3) is assumed to be a constant and economic material in many applications such as catalysis and sensors [49].

Furthermore, nano-hematite is also established to be an efficient adsorbent for the exclusion of heavy metals from normal water [50]. The 3D flowerlike α-Fe_2O_3 gathered from nanopetal blocks is produced for use in waste water treatment. Such smart structures could proficiently evade the supplementary accumulation and progress the surface area with various areas to interact with contaminants.

10.2.6 Carbon Nanotubes

This is a most successful kind of material, especially the varieties of carbon nanomaterial (CNMs), due to its outstanding and unique properties and wider application, including its sorption properties. Such materials are also beneficial because of their sorption assets such as fast kinetics, large surface area and selectivity [6]. Many CNMs such as carbon beads (CBs), nanoporous carbons (NPCs), CNTs, and many more are reported now and amongst them, CNTs rapidly improved and attracted researchers globally due to their unique properties [6]. Because of their exceptional properties such as large surface area and porous structures, CNTs have outstanding adsorption efficiencies against many contaminants, such as dichloro and ethylbenzene, Cu(II), Pb(II), Zn(II), Cd(II), and dyes. Based on their structural orientation, they are classified into two categories: (1) SWCNTs, comprise of single layer graphene sheet tubes, and (2) MWCNTs, comprise of multiple layers with a spacing of ≈0.34 nm (Figure 10.4). Both types of CNTs are in use to remove contaminants from water.

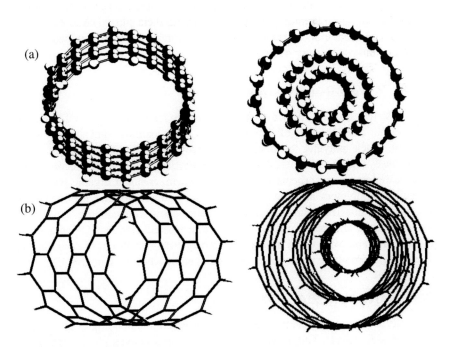

FIGURE 10.4 Structure of (a) SWCNTs and (b) MWCNTs.

CNTs can be combined with other metals to improve their optical, absorption, mechanical and electrical properties. Such functionalization increases the crowding on the CNTs surfaces which increases their surface areas and dispersibility [68 - 70]. For instance, in removing chromium from water, a combination of CNTs and Fe_2O_3 was prepared. An external magnetic field can easily eliminate the composite adsorbent from water.

10.2.7 Nanocomposites

There are many disadvantages such as aggregation, high bandgap, oxidation, and trouble in the separation of NMs which have been noticed and reported. Nanofiltration layers are also disturbed due to membrane fouling. CNTs are mostly constrained by their small capacity and high cost of production and due to the necessity for an associate medium. To overcome the above issues and to adopt a better way of increase effectiveness, nanocomposites are found to be the most suitable candidates for action on waste water. Recently, many studies have been reported, such as the chemical deposition of zero-valent metals on CNTs to develop a class of nanocomposites. The result indicates a good efficiency for nitrate removal from water. Because of the unique properties of composites, the adsorbent can be merely detached from the solution by applying a magnetic field. Moreover, the membranes of thin-film nanocomposites (TFN) are organized via in-situ TiO_2 NPs combined with a co-polyamide network fabrication. TiO_2 NPs efficiency can be improved by the incorporation of amine and chloride compounds. The NCs should be continuous with high nano-reactivity by altering their parent NMs. These are widely accepted materials due to their nontoxic natures, high stabilities, and are for fresh water and waste water treatment.

10.2.8 Polymer-Based Nanoadsorbents

In the past few decades, polymer nanocomposites have had increased attention towards the development of environmental sustainability and waste water treatment. Such polymer nanocomposites provide high surface areas, cost-effectiveness, higher stabilities, and high selectivities [51]. The most economical and used polymeric adsorbents are CS cyclodextrin, covalent organic polymers, extracellular polymers and the like [52]. Lately, lignin-derived nanomaterials have been synthesized and analyzed and have shown outstanding activity in dye degradation and heavy metal removal from waste water. A MOF of zeolitic imidazolate framework (ZIF) adsorbent, namely PES, which removes malachite green (MG) from waste water has been developed by Abdi and Abedini.

10.3 NANOMATERIAL TOXICITY

Previously, to a large degree, the waste water treatment methodologies gave many unwanted side effects. Chlorination is the most used conventional method for waste water treatment, but later it was observed that it generates many carcinogenic byproducts such as N-nitrosodimethylamine and trihalomethanes. Moreover, the need for nanotechnology in waste water treatment is now in focus. The properties of the nanomaterials are responsible for the resulting toxicity. Gold NPs below 6 nm enter the nucleus effectively.

In contrast, the bigger sizes 10 - 16 nm are only capable of puncturing the cell membrane and are therefore present in the cytoplasm. This was observed by Huo et al. The spherical shaped NPs are more inclined towards endocytosis than nanofibers and nanotubes. SWCNTs block calcium channels excellently in comparison with spherical shaped NPs. Nevertheless, fullerenes and CNTs exclude such calculations and establish the requirement for more toxicity investigations.

10.4 NANOMATERIAL ALTERATION HAZARD IN WATER

The interaction between bionanomaterials and their abiotic aspects, dispersibility or solubility, indicates the value of NMs. Normally, NPs settle down gradually as compared to bigger NPs, but due to their huge surface area, they adsorb additional residual particles due to their high-water insolubility.

10.4.1 Water Pollution

Many research groups work on toxicity-related issues caused due to nanomaterial-polluted water. The use of nanotechnology in the treatment of waste water is exposed to an organized examination of conceivable biological and ecotoxicity related to their use. The application of nanotechnology may give negative results for the above reasons, depending on what exactly is being examined by researchers who aim to decrease the expenses connected with NMs.

10.5 CONCLUSION

In this chapter, we focused mainly on the zero-valent metal NPs, metal oxide NPs, nanomaterials (NMs), nanocomposites (NCs), carbon nanotubes (CNTs), and polymer based nonabsorbents. Nanomaterial toxicity also discussed briefly, along with the major role of NMs in water pollution and its treatment. Such innovative NMs are discussed for their applications and roles in waste water treatment. The current need for research is based on those problems which affect livelihoods and due to this, these NMs look extremely promising for fresh water and waste water treatment. Previously, very limited classes of NMs have been reported commercially. Depending on their wider applications in waste water treatment, future research, and development needs to be dedicated to economic efficiency. Moreover, their wider application in water treatment raises concerns about their toxicity to humans and to environmental health because many studies have reported their adverse effects [1, 6, 7]. Additionally, very few studies are available on the toxicity of NMs, which needs to be explored extensively. Therefore, the periodic evaluation of their toxicity is a necessity in order to study the validities of real applications. Additionally, the uniformity in the performance and comparison of NMs in waste water treatment is the primary concern in current research to identify the promising ability of NMs. Therefore, the mechanism for their performance evaluation in waste water treatment must be high quality and uniform in the near future.

ACKNOWLEDGEMENTS

N.K.S. acknowledges the Science and Engineering Research Board, Government of India grant: SERB/CHE/2021412 for funding. We also express our sincere gratitude to Shri Maneklal M Patel Institute of Sciences and Research, Kadi Sarva Vishwavidyalaya, Gandhinagar, Gujarat, India.

REFERENCES

[1] Buzea, C. Pacheco, I. I. and Robbie, K. "Nanomaterials and nanoparticles: sources and toxicity."*Biointerphases* 2, no. 4 (2007), MR17–MR71. doi:10.1116/1.2815690.

[2] Parmon, V. "Nanomaterials in catalysis." *Materials Research Innovations* 12, no. 2(2008), 60–61. doi:10.1179/143307508x304228.

[3] Liang, X-J. Kumar, A. Shi, D. and Cui, D. "Nanostructures for medicine and pharmaceuticals. "*Journal of Nanomaterials* 2012 (2012), 1–2. doi:10.1155/2012/921897.

[4] Kusior, A. Klich-Kafel, J. Trenczek-Zajac, A. Swierczek, K. Radecka, M. and Zakrzewska, K. "TiO_2–SnO_2 nanomaterials for gas sensing and photocatalysis." *Journal of the European Ceramic Society* 33, no. 12 (2013), 2285–2290. doi:10.1016/j.jeurceramsoc.2013.01.022.

[5] Bujoli, B. Roussière, H. Montavon, G. Laïb, S. Janvier, P. Alonso, B. Fayon, F. et al. "Novel phosphate–phosphonate hybrid nanomaterials applied to biology." *Progress in Solid State Chemistry* 34, no. 2–4(2006), 257–266. doi:10.1016/j.progsolidstchem.2005.11.039.

[6] Khin, M. M. Nair, A. S. Babu, V. J. Murugan, R. and Ramakrishna, S. "A review on nanomaterials for environmental remediation." *Energy & Environmental Science* 5, no. 8 (2012), 8075. doi:10.1039/c2ee21818f.

[7] Tang, W-W. Zeng, G-M. Gong, J-L. Liang, J. Xu, P. Zhang, C. and Huang, B-B. "Impact of humic/fulvic acid on the removal of heavy metals from aqueous solutions using nanomaterials: a review." *Science of the Total Environment* 468–469 (2014), 1014–1027. doi:10.1016/j.scitotenv.2013.09.044.

[8] Yan, J. Han, L. Gao, W. Xue, S. and Chen, M. "Biochar supported nanoscale zerovalent iron composite used as persulfate activator for removing trichloroethylene." *Bioresource Technology* 175 (2015), 269–274. doi:10.1016/j.biortech.2014.10.103.

[9] Liu, F. Yang, J. H. Zuo, J. Ma, D. Gan, L. Xie, B. Wang, P. and Yang, B. "Graphene-supported nanoscale zero-valent iron: removal of phosphorus from aqueous solution and mechanistic study. "*Journal of Environmental Sciences* 26, no. 8 (2014), 1751–1762. doi:10.1016/j.jes.2014.06.016.

[10] Kalhapure, R. S. Sonawane, S. J. Sikwal, D. R. Jadhav, M. Rambharose, S. Mocktar, C. and Govender, T. "Solid lipid nanoparticles of clotrimazole silver complex: an efficient nano antibacterial against Staphylococcus aureus and MRSA." *Colloids and Surfaces B: Biointerfaces* 136 (2015), 651–658. doi:10.1016/j.colsurfb.2015.10.003.
[11] Borrego, B. Lorenzo, G. Mota-Morales, J. D. Almanza-Reyes, H. Mateos, F. López-Gil, E. De la Losa, N. et al. "Potential application of silver nanoparticles to control the infectivity of Rift Valley fever virus in vitro and in vivo." *Nanomedicine: Nanotechnology, Biology and Medicine* 12, no. 5 (2016), 1185–1192. doi:10.1016/j.nano.2016.01.021.
[12] Krishnaraj, C. Ramachandran, R. Mohan, K. and Kalaichelvan, P. T. "Optimization for rapid synthesis of silver nanoparticles and its effect on phytopathogenic fungi." *Spectrochimica Acta Part A: Molecular and Biomolecular Spectroscopy* 93 (2012), 95–99. doi:10.1016/j.saa.2012.03.002.
[13] Sondi, I. and Salopek-Sondi, B. "Silver nanoparticles as antimicrobial agent: a case study on E. coli as a model for Gram-negative bacteria." *Journal of Colloid and Interface Science* 275, no. 1 (2004), 177–182. doi:10.1016/j.jcis.2004.02.012.
[14] Danilczuk, M. Lund, A. Sadlo, J. Yamada, H. and Michalik, J. "Conduction electron spin resonance of small silver particles." *Spectrochimica Acta Part A: Molecular and Biomolecular Spectroscopy* 63, no. 1 (2006), 189–191. doi:10.1016/j.saa.2005.05.002.
[15] Dhanalekshmi, K. I. and Meena, K. S. "DNA intercalation studies and antimicrobial activity of Ag@ZrO_2 core–shell nanoparticles in vitro." *Materials Science and Engineering: C* 59 (2016), 1063–1068. doi:10.1016/j.msec.2015.11.027.
[16] Prabhu, S. and Poulose, E. K. "Silver nanoparticles: mechanism of antimicrobial action, synthesis, medical applications, and toxicity effects." *International Nano Letters* 2, no. 1 (2012). doi:10.1186/2228-5326-2-32.
[17] Li, X. Lenhart, J. J. and Walker, H. W. "Aggregation kinetics and dissolution of coated silver nanoparticles." *Langmuir* 28, no. 2 (2011), 1095–1104. doi:10.1021/la202328n.
[18] Quang, D. V. Sarawade, P. B. Jeon, S. J. Kim, S. H. Kim, J-K. Chai, Y. G. and Kim, H. T. "Effective water disinfection using silver nanoparticle containing silica beads." *Applied Surface Science* 266 (2013), 280–287. doi:10.1016/j.apsusc.2012.11.168.
[19] Ren, D. and Smith, J. A. "Retention and transport of silver nanoparticles in a ceramic porous medium used for point-of-use water treatment." *Environmental Science & Technology* 47, no. 8 (2013), 3825–3832. doi:10.1021/es4000752.
[20] Kallman, E. N. Oyanedel-Craver, V. A. and Smith, J. A. "Ceramic filters impregnated with silver nanoparticles for point-of-use water treatment in rural Guatemala." *Journal of Environmental Engineering* 137, no. 6 (2011), 407–415. doi:10.1061/(asce)ee.1943-7870.0000330.
[21] Oyanedel-Craver, V. A. and Smith, J. A. "Sustainable colloidal-silver-impregnated ceramic filter for point-of-use water treatment." *Environmental Science & Technology* 42, no. 3 (2007), 927–933. doi:10.1021/es071268u.
[22] Mikelonis, A. M. Youn, S. and Lawler, D. F. "DLVO approximation methods for predicting the attachment of silver nanoparticles to ceramic membranes." *Langmuir* 32, no. 7 (2016), 1723–1731. doi:10.1021/acs.langmuir.5b04675.

[23] Bratsch, S. G. "Standard electrode potentials and temperature coefficients in water at 298.15 K." *Journal of Physical and Chemical Reference Data* 18, no. 1 (1989), 1–21. doi:10.1063/1.555839.
[24] Bokare, V. Jung, J-L. Chang, Y-Y. and Chang, Y-S. "Reductive dechlorination of octachlorodibenzo-p-dioxin by nanosized zero-valent zinc: modeling of rate kinetics and congener profile." *Journal of Hazardous Materials* 250–251 (2013), 397–402. doi:10.1016/j.jhazmat.2013.02.020.
[25] Tratnyek, P. G. Salter, A. J. Nurmi, J. T. and Sarathy, V. "Environmental applications of zerovalent metals: iron vs. zinc." *ACS Symposium Series* (2010), 1045, 165–178. doi:10.1021/bk-2010-1045.ch009.
[26] Fujishima, A. and Honda, K. "Electrochemical photolysis of water at a semiconductor electrode." *Nature* 238, no. 5358 (1972), 37–38. doi:10.1038/238037a0.
[27] Guesh, K. Mayoral, Á. Márquez-Álvarez, C. Chebude, Y. and Díaz, I. "Enhanced photocatalytic activity of TiO_2 supported on zeolites tested in real waste waters from the textile industry of Ethiopia." *Microporous and Mesoporous Materials* 225 (2016), 88–97. doi:10.1016/j.micromeso.2015.12.001.
[28] Ohsaka, T. Shinozaki, K. Tsuruta, K. and Hirano, K. "Photo-electrochemical degradation of some chlorinated organic compounds on n-TiO_2 electrode." *Chemosphere* 73, no. 8 (2008), 1279–1283. doi:10.1016/j.chemosphere.2008.07.016.
[29] Guo, M. Song, W. Wang, T. Li, Y. Wang, X. and Du, X. "Phenyl-functionalization of titanium dioxide-nanosheets coating fabricated on a titanium wire for selective solid-phase microextraction of polycyclic aromatic hydrocarbons from environment water samples." *Talanta* 144 (2015), 998–1006. doi:10.1016/j.talanta.2015.07.064.
[30] Lee, Y-S. Kim, S-J. Venkateswaran, P. Jang, J-S. Kim, H. and Kim, J-G. "Anion co-doped titania for solar photocatalytic degradation of dyes." *Carbon Letters* 9, no. 2 (2008), 131–136. doi:10.5714/cl.2008.9.2.131.
[31] Nguyen, A. T. Hsieh, C-T. and Juang, R-S. "Substituent effects on photodegradation of phenols in binary mixtures by hybrid H2O2 and TiO2 suspensions under UV irradiation." *Journal of the Taiwan Institute of Chemical Engineers* 62 (2016), 68–75. doi:10.1016/j.jtice.2016.01.012.
[32] Gar Alalm, M. Tawfik, A. and Ookawara, S. "Comparison of solar TiO_2 photocatalysis and solar photo-Fenton for treatment of pesticides industry waste water: operational conditions, kinetics, and costs." *Journal of Water Process Engineering* 8 (2015), 55–63. doi:10.1016/j.jwpe.2015.09.007.
[33] Moon, G-H. Kim, D-H. Kim, H-I. Bokare, A.D. and Choi, W. "Platinum-such as behavior of reduced graphene oxide as a cocatalyst on TiO_2 for the efficient photocatalytic oxidation of arsenite." *Environmental Science & Technology Letters* 1, no. 2 (2014), 185–190. doi:10.1021/ez5000012.
[34] Kim, S. H. Lee, S. W. Lee, G. M. Lee, B-T. Yun, S-T. and Kim, S-O. "Monitoring of TiO_2-catalytic UV-LED photo-oxidation of cyanide contained in mine waste water and leachate." *Chemosphere* 143 (2016), 106–114. doi:10.1016/j.chemosphere.2015.07.006.
[35] Chen, Z. Li, Y. Guo, M. Xu, F. Wang, P. Du, Y. and Na, P. "One-pot synthesis of Mn-doped TiO_2 grown on graphene and the mechanism for removal of Cr(VI) and Cr(III)." *Journal of Hazardous Materials* 310 (2016), 188–198. doi:10.1016/j.jhazmat.2016.02.034.

[36] Foster, H. A. Ditta, I. B. Varghese, S. and Steele, A. "Photocatalytic disinfection using titanium dioxide: spectrum and mechanism of antimicrobial activity." *Applied Microbiology and Biotechnology* 90, no. 6 (2011), 1847–1868. doi:10.1007/s00253-011-3213-7.

[37] Anpo, M. Kishiguchi, S. Ichihashi, Y. Takeuchi, M. Yamashita, H. Ikeue, K. Morin, B. Davidson, A. and Che, M. "The design and development of second-generation titanium oxide photocatalysts able to operate under visible light irradiation by applying a metal ion-implantation method." *Research on Chemical Intermediates* 27, no. 4–5 (2001), 459–467. doi:10.1163/156856701104202101.

[38] Seery, M. K. George, R. Floris, P. and Pillai, S. C. "Silver doped titanium dioxide nanomaterials for enhanced visible light photocatalysis." *Journal of Photochemistry and Photobiology A: Chemistry* 189, no. 2–3 (2007), 258–263. doi:10.1016/j.jphotochem.2007.02.010.

[39] Page, K. Palgrave, R. G. Parkin, I. P. Wilson, M. Savin, S. L. and Chadwick, A. V. "Titania and silver–titania composite films on glass—potent antimicrobial coatings." *Journal Mater. Chem* 17, no. 1 (2007), 95–104. doi:10.1039/b611740f.

[40] Mohammadi, S. Harvey, A. and Boodhoo, K. V. "Synthesis of TiO_2 nanoparticles in a spinning disc reactor." *Chemical Engineering Journal* 258 (2014), 171–184. doi:10.1016/j.cej.2014.07.042.

[41] Chen, J. and Collier, C. P. "Noncovalent Functionalization of Single-Walled Carbon Nanotubes with Water-Soluble Porphyrins." *Journal. Phys. Chem.* B 2005, 109, 16, 7605–7609. . doi:10.1021/jp050389i.s001.

[42] Daneshvar, N. Salari, D. and Khataee, A. R. "Photocatalytic degradation of azo dye acid red 14 in water on ZnO as an alternative catalyst to TiO_2." *Journal of Photochemistry and Photobiology A: Chemistry* 162, no. 2–3 (2004), 317–322. doi:10.1016/s1010-6030(03)00378-2.

[43] Ge, F. Li, M-M. Ye, H. and Zhao, B-X. "Effective removal of heavy metal ions Cd^{2+}, Zn^{2+}, Pb^{2+}, Cu^{2+} from aqueous solution by polymer-modified magnetic nanoparticles." *Journal of Hazardous Materials* 211–212 (2012), 366–372. doi:10.1016/j.jhazmat.2011.12.013.

[44] Khaydarov, R. A. Khaydarov, R. R. and Gapurova, O. "Water purification from metal ions using carbon nanoparticle-conjugated polymer nanocomposites." *Water Research* 44, no. 6 (2010), 1927–1933. doi:10.1016/j.watres.2009.11.041.

[45] Liang, H. Xu, B. and Wang, Z. "Self-assembled 3D flower-such as α-Fe_2O_3 microstructures and their superior capability for heavy metal ion removal." *Materials Chemistry and Physics* 141, no. 2–3 (2013), 727–734. doi:10.1016/j.matchemphys.2013.05.070.

[46] Lu, C. Su, F. and Hu, S. "Surface modification of carbon nanotubes for enhancing BTEX adsorption from aqueous solutions." *Applied Surface Science* 254, no. 21 (2008), 7035–7041. doi:10.1016/j.apsusc.2008.05.282.

[47] Cho, H-H. Wepasnick, K. Smith, B. A. Bangash, F. K. Fairbrother, D. H. and Ball, W. P. "Sorption of Aqueous Zn[II] and Cd[II] by multiwall carbon nanotubes: the relative roles of oxygen-containing functional groups and graphenic carbon." *Langmuir* 26, no. 2 (2009), 967–981. doi:10.1021/la902440u.

[48] Li, Y-H. Ding, J. Luan, Z. Di, Z. Zhu, Y. Xu, C. Wu, D. and Wei, B. "Competitive adsorption of Pb^{2+}, Cu^{2+} and Cd^{2+} ions from aqueous solutions by multiwalled carbon nanotubes." *Carbon* 41, no. 14 (2003), 2787–2792. doi:10.1016/s0008-6223(03)00392-0.

[49] Madrakian, T. Afkhami, A. Ahmadi, M. and Bagheri, H. "Removal of some cationic dyes from aqueous solutions using magnetic-modified multi-walled carbon nanotubes." *Journal of Hazardous Materials* 196 (2011), 109–114. doi:10.1016/j.jhazmat.2011.08.078.

[50] Yang, C-M. Park, J. S. An, K. H. Lim, S. C. Seo, K. Kim, B. Park, K. A. Han, S. Park, C. Y. and Lee, Y. H. "Selective Removal of Metallic Single-Walled Carbon Nanotubes with Small Diameters by Using Nitric and Sulfuric Acids." Journal Phys. Chem. B *2005, 109, 41, 19242–19248*. doi:10.1021/jp053245c.s001.

[51] Soltani, T. and Lee, B-K. "Photocatalytic and photo-fenton catalytic degradation of organic pollutants by non-TiO_2 photocatalysts under visible light irradiation." *Current Developments in Photocatalysis and Photocatalytic Materials*, 2020, 267–284. doi:10.1016/b978-0-12-819000-5.00017-5.

[52] Zito, P. and Shipley, H. J. "Inorganic nano-adsorbents for the removal of heavy metals and arsenic: a review." *RSC Advances* 5, no. 38 (2015), 29885–29907. doi:10.1039/c5ra02714d.

Index

A

adsorbents 65–8, 75, 76
adsorption 64–80
aerogels, hydrogels 53
allotropes 37
antibacterial assay 90
antibiotics 26
antimicrobial 89–92

B

bacteria 186, 187, 195
ball milling 131
biodegradability 64, 65, 67, 73, 74
biodegradation 148
biofouling 155
bionanoparticles 84
biopolymers 63–5, 72, 74
bismuth nanoparticles 171
bottom-up 2
bovine serum albumin 34
brine 154, 155

C

carbon based nanomaterial 107
carbon nanotubes 4–6, 10, 14, 16, 17, 25, 26, 29–31, 33, 101, 103, 104, 170
cellulose 63–9, 71, 74, 76–9; acetate poly(vinylpyrrolidone) chiral 27, 31; based nanomaterials 63, 66, 67
ceramic 147, 153–5, 159
chemical oxygen demand 153
chemical vapor deposition 134
chitosan 63, 66, 68, 70–2, 74, 76–80, 84, 88, 89, 91, 92
COD 153, 154
contamination 64
copper nanoparticles 173
coprecipitation 156
crosslinked 70, 71, 73, 79, 80

D

decontamination 2, 3, 6, 7, 12, 15, 17
degradation mechanism of organic pollutants 128
desalination 148, 159
desorption 68, 72, 76
detection of chemical oxygen demand 173
detection of heavy metals (HMs) 168
detection of nitrates 167
detection of nutrients and phenolic compounds 166
detection of phenolic compounds 168
detection of phosphates 167
disc diffusion 90

E

ecofriendly 89
electrochemical, 27, 29, 31
electrocoagulation 148
electro-precipitation 64
electrostatic interaction 148
emerging organic contaminants 50
empty bed contact time (EBCT) 35
energy dispersive spectroscopy 158
ethylenediamine 151

F

fenton 148, 153, 154
FESEM 157, 158
flux recovery ratio 155, 157, 158
flux-selectivity 148
FTIR 157, 158
fullerenes 4, 6, 29, 101, 104, 105
functionalized 66, 72, 76

G

gold nanoparticles 170
graphene 4, 5, 9–11, 15, 17
graphene oxide (GO)/copper nanoparticles 175
graphite oxide 155

H

heavy metals 64, 65, 78, 80, 98, 99
hematite 9
hydrophilicity 148, 151, 153, 155
hydrophobic MOFs 55
hydrothermal 133

I

immiscible phases 148
immobilization 28, 64, 68
impregnation 132
infusion 29

interfacial polymerization 148
iron nanoparticles 7
iron oxides 172

M

macroporous 5
magnetic field 74
magnetic framework composites 56
magnetic MOF composites 49
malachite green 35
mechanical properties 67, 80
membrane fouling 148
membranes 65, 75, 77, 79
membrane separation 148, 157
mesoporous 5, 6
microbes 83, 87
microorganism 186–9
microporous 5
MOF 46–8
monovalent cations 152

N

nanophotocatalysts 137; carbonaceous 139; doping and co-doping 140; iron 140; MOFs 139; TiO_2 138; ZnO 137
nanoadsorbents 9, 16, 185, 192
nanocellulose 84, 88
nanocomposite 11, 13, 15, 26, 27, 37, 65, 66, 68–80, 108, 185, 186, 192, 193, 197
nanocrystalline 70, 76
nanofiltration membrane 147, 149, 150, 151, 153, 155–9
nanomembranes 3, 12, 14, 15
nanoparticles 155–7, 185–7, 189, 190, 194, 195, 197
nanophotocatalysts 12–15
nanoscaffold 151
nanosheets 155, 156, 159
nanosorbents 3, 9, 12, 15, 16
nanotechnology 74, 75, 186, 193, 195
nanotoxicology 185, 186
nanotubes 151, 157, 158, 185, 191, 193, 197, 198
natural organic matter 154
nickel nanoparticle, 34
nanomaterials 185, 186, 192–4, 197
NOM 154–7
non-biodegradability 65
NPs 155–7

O

organic nanocomposites 73

P

PA 148, 152
pathogens 82, 83, 87
pectin 63, 66, 70, 73, 74, 80
PEM 152, 153
permeate 152, 154, 155, 157, 158
PES 151, 157–8
photocatalytic nanomaterials 128, 131
photocatalytic 34, 37
photodegradation 148
photo-oxidation 148
physiological features 74
PIP 153
piperazine 153
plant based nanoparticles 89
polyamide 148, 151
polydopamine 156
polyelectrolyte multilayer 152
polyethersulfone 35, 151
polyhedral oligosilsesquioxane 158
polymer 185, 192, 193, 197
polysaccharides 63–6, 70, 72–4
polysulfone 152
pore dimensions 4
PSf 152, 155, 158–9
PSS 151, 153
pyrolysis 29

Q

quantum dots 3–5

R

reactive Black 5 (RB5) 30, 35
reactive oxygen species 8
remediation 2, 3, 5, 7, 8, 10, 12, 14–17
removal efficiencies 68
removal of dyes using MOFs 51
renewable 64, 66, 74, 80
reverse osmosis 148, 153
rhodamine 32, 34, 38

S

SEM 157
silica based nanoadsorbents 108
silver nanoparticles 89
single-walled carbon nanotubes (SWCNTs) 27, 39
sol-gel 131
solvothermal 133
starch 63, 66, 70–2, 74, 76, 80
sustainable 64, 74, 78, 80
synthesis methods 131

Index

T

temperatures 68
TFC 148, 151, 152, 159
TFN 153, 157
thermodynamic 28, 31
thin-film composite 148
TiO_2 NPs 90
TiO_2 particles 8, 9, 13
total organic carbon 153, 154
total suspended solids 154
triclosan (TCS) 28

U

UF 148, 154, 156, 159
ultrafiltration 148

W

wastewater 185–8, 189, 191–4, 196; treatment 64, 66, 67, 72, 74, 76, 78, 79
water pollution 82
water purification 147–51, 153, 155–9

X

x-ray photoelectron spectroscopy 157

Z

zeta potential 155
ZnO nanoparticles 90